Lecture Notes in Mathematics

A collection of informal reports and seminars
Edited by A. Dold, Heidelberg and B. Eckmann, Zürich

318

Recent Advances in Topological Dynamics

Proceedings of the Conference on Topological Dynamics, held at Yale University, June 19-23, 1972, in honor of Professor Gustav Arnold Hedlund on the occasion of his retirement.

Edited by Anatole Beck
University of Wisconsin, Madison, WI/USA

Springer-Verlag
Berlin · Heidelberg · New York 1973

AMS Subject Classifications (1970): 28A65, 34C35, 47A35, 54H20

ISBN 0-387-06187-8 Springer-Verlag Berlin · Heidelberg · New York
ISBN 0-387-06187-8 Springer-Verlag New York · Heidelberg · Berlin

Offsetdruck: Julius Beltz, Hemsbach/Bergstr.

Introduction

In June of 1972, on the occasion of the retirement of Professor G. A. Hedlund from Yale University, a conference was convened at Yale in the area of topological dynamics, the science which he did so much to develop. This volume represents some of the most recent work in this subject, some of it done during the conference itself (I can speak for certain about Corollary 4.3 and Paradox 4.6 of my own paper, but surely there are others as well). The occasion brought together dynamicists from as far as Israel, England and California under the aegis of the National Science Foundation, to compare notes, exchange ideas, and share the results of recent research efforts. By this volume, we share these now with "whom it may concern". In view of the growing importance of this field, we anticipate that there will be many. This volume of recent advances in topological dynamics is then our token of respect to the man who did so much to foster and build this field, Gustav Arnold Hedlund.

Anatole Beck

August 25, 1972

New Haven, Connecticut

Citation read to the Faculty of Graduate School of Yale University on the occasion of Professor Hedlund's retirement

Gustav Arnold Hedlund, Philip Schuyler Beebe Professor of Mathematics, was born on May 7, 1904, in Sommerville, Massachusetts. He holds an A.B. (1925) and Ph.D. (1930) from Harvard and an M.A. (1927) from Columbia. He has taught at Hunter College (1925-27), Bryn Mawr (1930-39), Virginia (1939-48) and Yale (1948-72). He served as Chairman of Mathematics Department at Yale for ten years (1949-59).

Hedlund's early research concerned the structure of geodesic flows. He was the first to prove the ergodicity of geodesic flows on closed surfaces of constant negative curvature and to establish mixing properties using horocycle flows. Next came his fundamental work with Marston Morse on symbolic dynamics followed by the development of topological dynamics with a former student, Walter Gottschalk, culminating in an AMS Colloquium Publication (1955). All of these topics remain vigorously active with much of the work being done by Hedlund's mathematical descendants.

At your retirement, Professor Hedlund, your colleagues honor you for your accomplishments as a mathematician, your skill and dedication as a teacher, your wisdom and integrity as a leader, and your generosity as a friend.

June 1972

Contents

F-EXPANSIONS REVISITED

Roy L. Adler*

§1 Introduction

Although we don't recognize the fact at the time, we're introduced at an early age to symbolic dynamics when taught about the decimal expansion of numbers. This representation can be viewed as a special case of the following scheme. With any mapping φ of a space X onto itself, but not necessarily one-to-one, and a finite or countable partition $\delta = \{I_1, I_2, \ldots,\}$ of X we have the opportunity of associating a symbolic sequence of integers to the points $x \in X$, namely

$$x \to (x_0, x_1, \ldots,)$$

where x_i are integers defined by

$$\varphi^i x \in I_{x_i} .$$

(If φ is one-to-one it becomes appropriate to assign a two sided sequence $(\ldots x_{-1}, x_0, x_1, \ldots)$ to x.) For the case of decimal expansion of points in the unit interval φx is taken to be the fractional part of 10x and $\delta = \{[\frac{k}{10}, \frac{k+1}{10}); k=0,1,\ldots,9\}$.

I would like to wander slightly away from the field of topological dynamics over into ergodic theory by taking up questions related to the frequency of occurrence of symbols in the above symbolic representation. For this we need measurability assumptions on φ and δ and most of all a φ-invariant probability measure on the measurable subsets of X which we shall denote by μ. We usually require that δ be a generator which means that distinct points have different representative sequences with the possible exception of some points in a set of measure zero. It follows from the ergodic theorem that in the representations of points outside an exceptional set of measure zero the frequency of a block $[i_0, \ldots, i_n]$ exists and

*This work was partially supported by the U. S. Air Force Office of Scientific Research under contract F44620-C-0063, P001.

equals $\mu(I_{i_0} \cap \varphi^{-1} I_{i_1} \cap \ldots \cap \varphi^{-n} I_{i_n})$. In the language of probability we can construct a stationary stochastic process from φ and δ by defining random variables $\xi_n(x) = x_n$ with the numbers $\mu(\bigcap_{k=0}^{n} \varphi^{-k} I_{i_k})$ as the statistics of the process. Using the notions of probability as a guide we can distinguish a variety of types of partitions from properties of these numbers. Two partitions α and β are said to be independent (with respect to μ) if $\mu(A \cap B) = \mu(A)\mu(B)$ for $A \epsilon \alpha$, and $B \epsilon \beta$. They are said to be ε-independent if $\sum_{\substack{A \epsilon \alpha \\ B \epsilon \beta}} |\mu(A \cap B) - \mu(A)\mu(B)| < \varepsilon$. In addition we make the following definitions: a partition δ is said to be Bernoulli (with respect to φ) if $\delta \vee \varphi^{-1} \delta \vee \ldots \vee \varphi^{-n+1} \delta$ and $\varphi^{-n} \delta \vee \ldots \vee \varphi^{-2n+1} \delta$ are independent for all n (here the symbol $\vee$ indicates the common refinement of partitions and $\varphi^{-k}\delta$ is defined by $\varphi^{-k}\delta = \{\varphi^{-k} I_1, \varphi^{-k} I_2, \ldots,\}$). A partition δ is said to be weak Bernoulli if for any $\varepsilon > 0$ there exists a k such that $\delta \vee \ldots \vee \varphi^{-n+1}\delta$ and $\varphi^{-n+1-k} \delta \vee \ldots \vee \varphi^{-2n+1-k} \delta$ are ε-independent for all n. We can now state a consequence of the theorem of Friedman and Ornstein (2).

If a measure preserving transformation has a weak Bernoulli partition δ then for any invertible measure preserving transformation σ having Bernoulli partition β for which the number $-\sum_{B \epsilon \beta} \mu_\sigma(B) \log \mu_\sigma(B)$ is sufficiently large there exists a partition α of the domain of σ such that the (σ, α) process has the same statistics as the (φ, δ)-process. We shall comment on this further in connection with the main theorem.

§2 f-Expansions of Points in the Unit Interval

Let $X = [0, 1)$, λ denote Lebesgue measure, and $\delta = \{I_1, I_2, \ldots,\}$ be a finite or countable partition of X into half open intervals. Let φ be a mapping of X onto itself such that φ maps each element I of δ one-to-one onto X. The sequences $(x_0, x_1, \ldots,)$ given by $\varphi^i x \epsilon I_{x_i}$ are the so-called f-expansions of points of the unit interval. The term "f-expansion" was originally used because the expansion was expressed in the form $x = f(x_0 + f(x_1 + f(\ldots)))$ where f is a function related to φ but not to be confused with it. Continued fractions $x = 1/(x_0 + 1/(x_1 + \ldots))$ fit the (φ, δ)-scheme by setting φx equal to the fractional part of $1/x$ and $\delta = \{[\frac{1}{k+1}, \frac{1}{k}) ; k = 1, 2, \ldots\}$. It can be shown by direct verification that $d\mu = dx/(1+x)\log 2$ is a φ-invariant probability measure. This is the famous Gauss measure from which the frequency of occurrence of integers in continued fraction expansions can be calculated. This fact and other independence properties of blocks of symbols in the expansion follow from a general theorem.

Theorem If

(i) φ is twice continuously differentiable on each $I \in \delta$,

(ii) there exists an integer ν such that $\inf\limits_{\substack{x \in I \\ I \in \delta}} \left| \frac{d}{dx} \varphi^{\nu}(x) \right| > 1$,

(iii) $\sup\limits_{x, y, z \in I} \left| \varphi''(x)/\varphi'(y)\varphi'(z) \right| < \infty$,

then

(iv) δ is a generator,

(v) $\mu(E) = \lim \lambda(\varphi^{-n}E)$ exists and defines a unique φ-invariant ergodic probability measure equivalent to Lebesgue measure,

(vi) δ is weak Bernoulli with respect to μ. In fact $\delta^{(n)}$ and $\varphi^{-n-k}\delta^{(n)}$ are $c\theta^{\sqrt{k}}$-independent where $0 < \theta < 1$ and $\delta^{(n)}$ denotes $\delta \vee \varphi^{-1}\delta \vee \ldots \vee \varphi^{-n+\delta}$.

We can understand hypotheses (ii) and (iii) in the following manner. Hypothesis (ii) is an expansiveness condition by means of which the identity transformation and any mapping coinciding with it on a set of positive measure are eliminated, a sine qua non for ergodicity. Hypothesis (iii) can be viewed in the following light. If φ were linear on each I, then Lebesgue measure itself would be the ergodic φ-invariant measure and δ would be Bernoulli. For such a map the quantity in (iii) vanishes; so (iii) is a condition limiting the departure from linearity. The proof depends on estimates of the numbers $M_n \equiv \sup\limits_{\substack{x, y \in I \\ I \in \delta^{(n)}}} \left| \frac{d}{dx}\varphi^n(x) / \frac{d}{dx}\varphi^n(y) \right|$. These are so important because of the fact that

$$1/M_n \leq \lambda(\varphi^{-n}E \cap I)/\lambda(E)\lambda(I) \leq M_n$$

for $I \in \delta^{(n)}$. The quantities that appear in (ii) and (iii) play an important role in showing that M_n is bounded away from 0 and ∞. It should be appreciated that the hypotheses of the above theorem are of the type easily checked for most examples. The numerous authors who have written on this material usually do not present a theorem with this in mind. For the continued fraction mapping we have $\inf\left|\frac{d\varphi^2}{dx}\right| = 4$ and $\sup|\varphi''(x)/\varphi'(y)\varphi'(z)| = 16$. The proof of the theorem is contained essentially in the work of Renyi (3) and Doeblin (1). There is one feature of the proof that has always amused me. The quantity $\mu(E)$ is initially defined as a generalized limit of $\lambda(\varphi^{-n}E)$. It then turns out using the methods of ergodic theory that the limit actually exists. The convergence is even exponential. The proof of (vi) is by far the hardest part of the theorem and follows closely the one of Doeblin (1) for the stochastic process associated with continued fraction expansions. Doeblin used conclusion (vi) to obtain a central limit theorem and law of the iterated logarithm for this process. This property is also the same one needed to apply the result of Friedman and Ornstein which then states that the process that produces digits in expansions, like continued fractions, is statistically the same as one that can be achieved with a roulette wheel by considering some other complicated combination of outcomes as events rather than the familiar simple independent ones. The requirements on the roulette wheel are that it have enough slots and have been forever and ever in play!

§ 3 Infinite Invariant Measure

An interesting situation occurs if one tries to generalize this theorem by relaxing hypothesis (ii). If φ is allowed to have a fixed point x_0 at which $\varphi'(x_0) = 1$, then the following examples reveal that an infinite invariant ergodic Lebesgue equivalent measure exists rather than a finite one.

<u>Example 1.</u> $\varphi(x)$ is defined as the fractional part of $x/1-x$ and $\delta = \{[\frac{k}{k+1}, \frac{k+1}{k+2}): k=0,1,\ldots\}$ in which case $\varphi(0) = 0$ and $\varphi'(0) = 1$. I learned from W. Parry that for this mapping Renyi showed in the Hungarian version of (3) that $d\mu = dx/x$ is an infinite invariant ergodic measure.

<u>Example 2.</u> Letting $x = [0, 2\pi)$ define φ by

$$\varphi(x) = \begin{cases} 2 \arctan (1/2) \tan x, & 0 \le x < \pi \\ 2 \arctan (1/2) \tan (x-\pi), & \pi \le x < 2\pi . \end{cases}$$

Again $\varphi(0) = 0$ and $\varphi'(0) = 1$. For this mapping $d\mu = dx/1-\cos x$ is an infinite invariant measure although the details of the proof of its ergodicity are still to be worked out. In both of these examples we have a remarkable stroke of good fortune similar to the case of Gauss measure where a measure exists whose form is simple and whose invariance can be verified directly. In conclusion I believe there is a general theorem lurking here which will explain the dramatic shift from finite to infinite measure exhibited in the above examples.

Finally I would like to mention that I gradually became acquainted with this area over the years from various conversations, originally with R. Scoville who independently discovered Renyi's theorem on f-expansions circa 1960, but mostly with L. Flatto and B. Weiss several years later.

REFERENCES

[1] Doeblin, W., Remarques sur la theorie metrique des fractions continues, Compositio Math. 7 (1940) 353-371.

[2] Friedman, N. and Ornstein, D., On isomorphism of weak Bernoulli transformations, Advances in Math. 5 (1970) 365-394.

[3] Renyi, A., Representations for real numbers and their ergodic properties, Acta Math. Akad. Sci. Hungar. 8 (1957) 477-493.

International Business Machines Corporation
Thomas J. Watson Research Center
Yorktown Heights, New York

NON-COMPACT DYNAMICAL SYSTEMS

Joseph Auslander
University of Maryland
University of California

In this paper, two topics in the theory of non-compact dynamical systems are considered. The first of these is a generalization of earlier work on Liapunov stability of compact invariant sets. The second concerns a situation , the absence of recurrence, which has no compact counterpart.

Suppose a dynamical system $(x,t) \to xt$ is given on the locally compact metric space $X(x \in X,\ t \in \mathbb{R} = (-\infty,\infty))$. We write $\gamma(x)$ and $\gamma^+(x)$ for the orbit and positive semiorbit, respectively, of x. We recall the definitions of prolongations and prolongational limit sets ([4],[2]). The first prolongation of x, $D_1(x) = \bigcap \left[\overline{\gamma^+(U)} \mid U \in \eta_x\right]$, where η_x denotes the neighborhood system of x. We may also write

$$D_1(x) = \left[y \mid x_n \to x, t_n \geqslant 0, x_n t_n \to y, \text{ for some sequences } \{x_n\} \text{ in } X, t_n \geqslant 0\right]$$

The higher prolongations are defined inductively: if α is an ordinal number, and $D_\beta(x)$ has been defined, for $\beta < \alpha$, then

$$D_\alpha(x) = \left[y \,\middle|\, \begin{array}{l} \text{there are sequences } x_n \to x,\ y_n \to y,\ \beta_n < \alpha \\ \text{with } y_n \in D_{\beta_n}^{k_n}(x_n) \quad (k_n > 0) \end{array}\right].$$

(If $P\colon X \to 2^X$ is a set function, then

$$P^k(x) = [y \mid \text{ there are } x_0 = x, x_1, \ldots, x_k = y, \text{ with } x_i \in P(x_{i-1})].)$$

The prolongational limit sets $\Lambda_\alpha(x)$ are subsets of the $D_\alpha(x)$: $\Lambda_1(x) = \left[y \mid x_n \to x, t_n \to \infty, x_n t_n \to y, \text{ for some } x_n, t_n\right]$,

and, inductively, $$\Lambda_\alpha(x) = \left[y \,\middle|\, \begin{array}{l} x_n \to x,\ y_n \to y,\ y_n \in \Lambda_{\beta_n}^{k_n}(x_n), \\ \text{for sequences } x_n, y_n,\ \beta_n < \alpha,\ k_n > 0 \end{array}\right].$$

The $\{D_\alpha(x)\}$ and $\{\Lambda_\alpha(x)\}$ are increasing families of closed sets, and, for each $x \in X$, $D_\alpha(x) = \gamma^+(x) \cup \Lambda_\alpha(x)$.

The prolongations have proved to be useful in the study of stability of compact invariant sets. If M is such a set, it is said to be <u>Liapunov stable</u>, if, for any neighborhood W of M, there is a neighborhood U of M, with $\gamma^+(U) \subset W$; equivalently, if $\varepsilon > 0$, there is a $\delta > 0$ such that if $d(x,M) < \delta$, then $d(xt,M) < \varepsilon$, for $t \geqslant 0$. By a theorem of Ura, [8], M is Liapunov stable if and only if $D_1(M) = M$. The higher prolongations are used to define more general stability concepts: M is <u>stable of order</u> α if $D_\alpha(M) = M$ (so stability of order 1 is Liapunov stability.)

Liapunov functions are another tool in the study of stability. These are non-negative functions such that $V^{-1}(0) = M$, $V(xt) \leqslant V(x)$ $(x \in X,\ t > 0)$ and such that $V(x_n) \to 0$ if and only if $x_n \to M$. Then M is Liapunov stable if and only if there exists a Liapunov function, and stable of order α, for every α, ("absolutely stable") if and only if there is a continuous Liapunov function, [4].

Now suppose M is a closed (but not compact) invariant subset of X. Note that the two definitions given above of Liapunov stability (the "neighborhood" and the "ε - δ" definitions) are not equivalent in this case. Rather than choose one of them as the general definition of Liapunov stability, we proceed as follows. Let $\mathcal{N} = \mathcal{N}_M$ denote the neighborhood system of M, and let $\mathcal{F}$ be a neighborhood filter of M (that is, $\mathcal{F}$ is a filter on X with $\mathcal{F} \subset \mathcal{N}_M$.) Write $\mathcal{F}^+$ for the filter generated by the positively invariant members of $\mathcal{F}$. Then always $\mathcal{F}^+ \subset \mathcal{F}$, and M is said to be $\mathcal{F}$-stable if $\mathcal{F}^+ = \mathcal{F}$. Then M is $\mathcal{F}$ stable if and only if, whenever $F \in \mathcal{F}$, there is an $F' \in \mathcal{F}$ such that $\gamma^+(F') \subset F$. Thus each neighborhood filter gives rise to a stability notion-important special cases are <u>topological</u>

stability ($\mathcal{F} = \eta_M$) and metric stability ($\mathcal{F}$ = the filter generated by the metric neighborhoods of M).

Now, M is always $\mathcal{F}$ stable, for some neighborhood filter $\mathcal{F}$. Indeed, if $\mathcal{G}$ is any neighborhood filter, then obviously M is $\mathcal{G}^+$ stable. In order to exclude such trivial cases, we consider neighborhood filters with the additional property that $\bigcap[\overline{F} | F \in \mathcal{F}] = M$ (we call such filters sufficient).

We define the $\mathcal{F}$-prolongation of M, $D_{\mathcal{F}}(M) = \bigcap[\overline{\gamma^+(F)} | F \in \mathcal{F}]$. The following theorem connects filter stability with prolongations.

Theorem. Let M be a closed invariant set. Then the following are equivalent:

(i) M is $\mathcal{G}$ stable, for some sufficient neighborhood filter $\mathcal{G}$.

(ii) $D_{\mathcal{F}}(M) = M$ for some neighborhood filter $\mathcal{F}$.

(iii) $D_1(x) \subset M$, for all $x \in M$.

Topological stability is closely connected with compactness, as the next theorem indicates (for related results see [6]).

Theorem. Let M be a closed invariant set with no interior. Then the following are equivalent:

(i) M is topologically stable.

(ii) $D_1'(x)$ is a compact subset of M, for every $x \in M$.

(iii) For each $x \in X$, $D_1(x)$ is compact, and M is $\mathcal{F}$ stable, for some sufficient filter $\mathcal{F}$.

Similar definitions and theorems may be formulated for asymptotic stability, [3].

Just as in the compact case, Liapunov function can be used to study stability. However, stability cannot be determined by a single Liapunov function (since, in general, a neighborhood filter $\mathcal{F}$ does not have a countable base).

If M is a closed invariant set, and $\mathcal{F}$ a neighborhood filter, a real valued function V will be called a <u>weak</u> $\mathcal{F}$ <u>Liapunov function</u> if (i) $V \geq 0$ and $V^{-1}(0) = M$, (ii) $V(xt) \leq V(x)$, for $x \in X$, $t > 0$, and (iii) $V^{-1}([0,\varepsilon)) \in \mathcal{F}$, for all $\varepsilon > 0$. If $\mathcal{V}$ is a collection of weak Liapunov functions it will be called an $\mathcal{F}$ <u>Liapunov family</u> (for M) provided that (iv): for every $F \in \mathcal{F}$, there is a $V \in \mathcal{V}$ such that $\inf [V(x) \mid x \notin F] > 0$.

The properties of $\mathcal{F}$ Liapunov families can be paraphrased as follows. Let $\{x_n\}$ be a net in X; we write $x_n \overset{\mathcal{F}}{\to} M$ if $\{x_n\}$ is eventually in every $F \in \mathcal{F}$. If $\mathcal{V}$ is an $\mathcal{F}$-Liapunov family, then $x_n \overset{\mathcal{F}}{\to} M$ if and only if $V(x_n) \to 0$, for all $V \in \mathcal{V}$.

Using Liapunov families, the classical stability theorem can be generalized.

<u>Theorem</u>. Let M be a closed invariant subset of the second countable locally compact metric space X and let $\mathcal{F}$ be a sufficient neighborhood filter of M. Then M is $\mathcal{F}$ stable if and only if there is an $\mathcal{F}$-Liapunov family for M.

Our second topic is the study of dynamical systems which have no generalized recurrent points. A point $x \in X$ is said to be <u>recurrent of order</u> α, if $x \in \Lambda_\alpha(x)$. We write $\mathcal{R}_\alpha$ for the set of such points, and call $\mathcal{R} = \bigcup_\alpha \mathcal{R}_\alpha$ the generalized recurrent set. $\mathcal{R}$ includes the Poisson stable and non-wandering points. A compact dynamical system always includes recurrent points (in fact, almost periodic points) so a dynamical system for which $\mathcal{R}$ is empty is an essentially non-compact phenomenon. We call such systems <u>gradient dynamical systems</u>; they are characterized by the existence of a continuous real valued function f such that $f(xt) < f(x)$, for all $x \in X$ and $t > 0$, [2].

Now, if $\Lambda_1(x) = \phi$, for all $x \in X$, then it follows easily that $\Lambda_\alpha(x) = \phi$ for all α and all x, and the flow is clearly gradient. In this case, the dynamical system is known to be <u>parallelizable</u>—there is a closed set S in X such that X is homeomorphic to $S \times \mathbb{R}$, and the flow corresponds to translation in the second coordinate, [1]. The general case (gradient, but not parallelizable) is characterized by the existence of orbits which have non-empty prolongational limit sets. These orbits are separatrices, as defined by Markus [7]—regarded as elements of the orbit space $\tilde{X}$, they are the points where $\tilde{X}$ fails to be Hausdorff, or limits of such points.

Let f be a function decreasing along every orbit, as described above. We may suppose the range of f is $J = (-1,1)$. Then if $X_r = f^{-1}(r)$ and $\Sigma_r = [x \mid xt \in X_r,$ for some $t]$ $(r \in J)$, Σ_r is an invariant set, and the flow restricted to it is obviously parallelizable. Hence a gradient flow contains parallelizable subflows. Conversely, it is possible to synthesize a gradient flow from a given family of parallel flows.

<u>Theorem</u>. Let $X_r, (r \in J)$ be locally compact, second countable metric spaces. For $r, s \in J$, let $X_{r,s}$ be open subsets of X_r, with $X_{r,r} = X$, and such that $X_{r,s} \uparrow X_r$ as $s \to r^+$ and $s \to r^-$.

Suppose there are homeomorphisms $\phi_{r,s}: X_{r,s} \to X_{s,r}$ such that $\phi_{r,r}$ = identity and such that $\phi_{s,\tau}\phi_{r,s} = \phi_{r,\tau}$ (in the sense that if the left side is defined, so is the right side and they are equal, and if $r > s > \tau$ and $\phi_{r,\tau}(x)$ is defined, so are $\phi_{r,s}(x)$ and $\phi_{s,\tau}\phi_{r,s}(x)$, and both sides are equal; similarly if $r < s < \tau$).

Also, suppose if $x \in X_r$, $\left[s \mid x \in X_{r,s}\right]$ is an open set (hence an interval containing r).

Let $X = \bigcup_{r \in J} X_r$ be topologized by defining a basic neighborhood (in X) of $x \in X_r$ to be $N_{r,\varepsilon} = \bigcup\left[\phi_{r,s}(N) \mid s \in (r-\varepsilon, r+\varepsilon)\right]$, where

N is a basic neighborhood of x in X_r. Then X is a second countable locally compact metric space, and there is a gradient flow on X. Moreover, all gradient dynamical systems may be obtained in this manner.

The homeomorphisms $\phi_{r,s}$ are used to define a "partial flow" on X - that is, xt is defined for t in an open interval (α_x, β_x). Using a method due to Beck [5], the orbits are reparameterized, and a flow in the usual sense is obtained.

REFERENCES

1. H. A. Antosiewicz and J. Dugundji, Parallelizable flows and Liapunov's second method, Ann. Math, 73(1961), 543-555.
2. J. Auslander, Generalized recurrence in dynamical systems, Contributions to Differential Equations, III (1964), 65-74.
3. J. Auslander, On stability of closed sets in dynamical systems, Springer Lecture Notes in Mathematics, Vol. 144, (1970), 1-4.
4. J. Auslander and P. Seibert, Prolongations and stability in dynamical systems, Ann. Inst. Fourier, XIV, (1964), 237-268.
5. A. Beck, On invariant sets, Ann. Math, 67(1958), 99-103.
6. O. Hajek, Compactness and asymptotic stability, Math. Systems Theory 4(1970), 154-156.
7. L. Markus, Parallel dynamical systems, Topology, 8(1969), 47-57.
8. T. Ura, Sur le courant exterieur à une region invariante, Funke. Ekvac. 2(1959), 143-200.

ERGODIC G-INDUCED FLOWS ON COMPACT SOLVMANIFOLDS

Louis Auslander*

Institute for Advanced Study

and

Graduate Center, C.U.N.Y.

1. Introduction

In [1] and two subsequent papers [2] and [3], L. Green and the author developed an algorithm for determining when a G-induced flow on a compact solvmanifold is ergodic. In this paper we will use the algebraic results developed in [4] to connect this algorithm into necessary and sufficient conditions for a G-induced flow to be ergodic on a compact solvmanifold. We will denote the rest of this section to reviewing notation and language from [2] and [4]. Section 2 contains a survey of what was achieved in [1], [2] and [3]. Our main results are stated in theorems A and B of section 3. In presenting the details of this paper we will assume that the reader has some familiarity with [4] and the facts about G-induced flows on nilmanifolds contained in [1].

Let R/D be a presentation of a compact solvmanifold; namely, R is a connected simply connected solvable Lie group, D is a closed subgroup of R which contains no non-trivial connected subgroup normal in R, and R/D is compact. Then the fundamental group of R/D, Γ, is isomorphic to D/D_o, where a subscript zero will always denote the identity component of the group. Let $p(\xi), \xi \in \mathbb{R}$, be a

* John Simon Guggenheim Fellow and partially supported by grant from National Science Foundation.

one parameter subgroup of R and consider $p(\xi)$ as a group of transformations acting on R/D. This action of $p(\xi)$ on R/D will be called a <u>G-induced flow</u> (G-induced is an abbreviation for group induced) on R/D. If N is the nil-radical of R, we will say that $p(\xi)$ is in <u>general position</u> if $p(\xi)$ induces an ergodic action on the torus R/DN, where DN stands for the subgroup of R generated by D and N.

As a matter of general notation we will adopt the notation $A \rtimes B$ or $B \ltimes A$ to denote the semi-direct product of A and B, where A is assumed to be normal.

We will now introduce some semi-simple splittings in a systematic way. Let $R_S = M_R \rtimes T_R$ be the semi-simple splitting of R and let the semi-direct product representation of R_S be chosen so that if $p(\xi) = m(\xi)\, t(\xi)$, $m(\xi) \subset M_R$ and $t(\xi) \subset T_R$, then $m(\xi)t(\xi) = t(\xi)\, m(\xi)$ and each is a one parameter group. <u>We will call $m(\xi)$ the unipotent part of $p(\xi)$ and $t(\xi)$ the semi-simple part of $p(\xi)$.</u> Let M be the algebraic hull of the nil-shadow of Γ and form $M \rtimes T$, where T is the semi-simple part of Γ. Then there exists a homomorphism

$$\rho : M_R \longrightarrow M$$

with kernel D_o.

Now let $A_h(T)$ denote the algebraic hull of T in the group of automorphisms of M and let $A_h(T)_o$ be the identity component of $A_h(T)$. Let Γ^* be the maximal subgroup of finite index in Γ such that the algebraic hull of the semi-simple parts of Γ^* is connected. We will denote the semi-simple parts of the group Γ^* by $T^* \subset T$.

Let us now recall the concept of the <u>unstable ideal</u> as introduced in [2]. Let $\mathcal{L}$ be a Lie algebra and let $X \in \mathcal{L}$. Decompose $\mathcal{L}$ as a vector space into primary components relative to ad X:

$$\mathcal{L} = \mathcal{L}_o \oplus \mathcal{L}_1 \oplus \ldots \oplus \mathcal{L}_m$$

Here each $\mathcal{L}_i$ is invariant under ad X, and the minimal polynomial of the restriction of ad X to $\mathcal{L}_i$ is an irreducible polynomial $\mu_i(\lambda)$. $\mathcal{L}_i$ is the set of elements Z in $\mathcal{L}$ such that $\mu_i(\text{ad } X)^r Z = 0$ for some r. We reserve the index 0 for the null component of ad X, namely, $\mu_0(\lambda) = \lambda$. We will call $\mathcal{L}_i$ an unstable component if the roots of $\mu_i(\lambda) = 0$ have non-vanishing real parts. The smallest subalgebra containing all the unstable components is an ideal will be called the unstable ideal relative to X.

If $p(\xi)$ is a one parameter subgroup of R we will define the unstable normal subgroup U relative to $p(\xi)$ by passing to the Lie algebra of R, L(R), and, letting X denote the tangent vector to $p(\xi)$ at the origin, define U as the connected subgroup of R whose Lie algebra is the unstable ideal relative to X. It is a simple consequence of the Jordan canonical form theorem that $p(\xi)$ and its semi-simple part $t(\xi)$ have the same unstable normal subgroup.

2. Resumé and Augmentation of results from [1], [2] and [3].

In this section we will organize and amplify the results from [1], [2] and [3] so as to make them readily available for our needs in the next section.

An immediate application of the unstable ideal is the following generalized Mautner Lemma from [2]

Theorem 1. Let $g \longrightarrow U(g)$ be a continuous unitary representation of a connected Lie group whose Lie algebra is $\mathcal{L}$. If

$$U(\exp \xi X)\Psi = e^{i\lambda}\Psi \qquad \text{some } \lambda \in \mathbb{C} \text{ and } \xi \in \mathbb{R}$$

Then $U(\exp Y)\Psi = \Psi$ for every Y in the ideal of $\mathcal{L}$ unstable relative to X.

Theorem 1 was then strengthened in [2] by proving the following result.

Theorem 2. Let $p(\xi)$ be a G-induced flow on R/D and let U be the unstable normal subgroup relative to $p(\xi)$. Let $\phi: R/D \to R/(DU)^-$, where the bar denotes the closure be the canonical mapping. Then every eigen-vector of $p(\xi)$ acting on $L^2(R/D)$ is of the form $\Psi \cdot \phi$, where Ψ is an eigen vector of $p(\xi)$ acting on $L^2(R/(DU)^-)$.

(Remark: If we had been in possession of Theorem 2 when L. Green wrote Chapter VII of [1] The Appendix would not have been necessary.)

Theorem 2 has the following immediate corollary.

Corollary. Let $p(\xi)$ be a G-induced flow on R/D and let U be the unstable normal subgroup relative to $p(\xi)$. Then $p(\xi)$ acts ergodically on R/D if and only if $p(\xi)$ acts ergodically on $(R/U)/((UD)^-/U)$.

In [3] we began the study of R/U and proved the following result which we should have stated in the following slightly stronger form, as we actually proved this slightly stronger result.

Theorem 3. Let $p(\xi)$ be a G-induced flow on R/D which is in general position and has unstable normal subgroup U. Let H be the maximal analytic subgroup of $(DU)^-$ normal in R. Then R/H is class R.

(Recall that a class R solvable group is one when each element on T_R has all eigen values of absolute value one.)

We must now return to [1] for a study of class R solvmanifolds, i.e., solvmanifolds of the form R/D where R is a class R solvable Lie group.

Let us begin by reviewing some algebraic results from [4]. Let R/D be a presentation of a class R solvmanifold. If

$R_S = M_R \rtimes T_R$ is the semi-simple splitting of R it follows from [4] that $R/D \cap M_R$ is compact, T_R is compact, and that the projection

$$p: R \longrightarrow M_R$$

induces a homeomorphism $p^*: R/D \cap M_R \longrightarrow M_R/D \cap M_R$. Hence R/D is finitely covered by the compact nilmanifold $M_R/D \cap M_R$ which is homeomorphic to $(M_R/D_o)/(D \cap M_R/D_o)$. (Recall D_o is normal in M_R.)

Let us now state and prove a slightly more general result from Theorem 4.2, the main non-algebraic result of Chapter VI of [1].

Theorem 4. Let $p(\xi)$ be a G-induced flow in general position on the class R solvmanifold R/D. Then the flow $p(\xi)$ on $R/D \cap M_R$ is equivalent to the unipotent part of $p(\xi)$, $m(\xi)$, acting on $M_R/D \cap M_R$.

Proof: One knows that $m(\xi_1)\, t(\xi_2) = t(\xi_2)\, m(\xi_1), \xi_1, \xi_2 \in R$, where $m(\xi)$, $t(\xi)$ are the unipotent and semi-simple parts of $p(\xi)$, respectively. Since $p(\xi)$ is in general position, $t(\xi)$ is dense in T_R. Hence T_R acts trivially on $m(\xi)$. Let

$$p^*: R/D \cap M_R \longrightarrow M_R/D \cap M_R$$

be the homomorphism defined above. Since T_R acts trivially on $m(\xi)$ it is immediate that

$$p^* \circ p(\xi) = m(\xi) \circ p^* \qquad \xi \in \mathbb{R}$$

This proves our assertion. (For a more general result see Lemma 1 of [3].)

Let us now recall two general facts. Let X and Y be two compact manifolds, let $p: X \longrightarrow Y$ be a covering mapping, and let $x(\xi)$ be a flow on X and $y(\xi)$ a flow on Y such that $p \circ x(\xi) = y(\xi) \circ p$. Then $x(\xi)$ is distal (minimal) if and only if $y(\xi)$ is distal (minimal).

We now come to a theorem that is very important in our future work.

Theorem 5. Let $p(\xi)$ be a G-induced flow in general position on the presentation R/D of a compact solvmanifold. Further, let $D_1 \subset R$ be commensurable with D. Then $p(\xi)$ acts ergodically on R/D if and only if it acts ergodically on R/D_1.

Proof: Since the unstable normal subgroup U of R relative to $p(\xi)$, does not depend on D, and D and D_1 are commensurable, it follows easily that $(UD)_o^- = (UD_1)_o^-$. Hence the maximal analytic subgroup H of $(UD)^-$ normal in R has the same property relative to $(UD_1)^-$. Thus $p(\xi)$ acts ergodically on R/D if and only if it acts ergodically on $(R/H)/(HD/H)$ and this statement is also true with D_1 replacing D. But since for class R solvmanifolds $p(\xi)$ acts ergodically if and only if it does so for all finite coverings, it follows that $p(\xi)$ acts ergodically on $(R/H)/(HD/H)$ if and only if it acts ergodically on $(R/H)/(HD/H)$. This proves our assertion.

3. Main Results

Using Theorem 5 and the notation from section 1, we may henceforth assume that $T = T \cap A_h(T)_o$ or $T^* = T$. Before going on let us take this opportunity to remark on two facts about semi-simple splittings. First, if I is a connected normal subgroup of R then I is normal in R_S the semi-simple splitting of R. Second, if D is cocompact in R and the semi-simple splitting $D_S = M_D \rtimes T_D$ we have $M_D \subset M_R$ and we may choose $T_D \subset T_R$. Further M_R is the algebraic hull of M_D.

Theorem 6. Let R be a solvable Lie group and let D be a closed cocompact subgroup of R such that $T_R \subset A_h(T_D)$ using the above notation. Further, let I be a connected normal subgroup of R.

Then $(DI)_o^-$ is normal in R.

Proof: We will prove that $(DI)_o^-$ is normal in R_S and hence it is normal in R. We first observe that since $(DI)_o^-$ is invariant under D it is invariant under D_S. Hence $(DI)_o^-$ is invariant under M_D. But since M_D is Zarisky dense in M_R, $(DI)_o^-$ is invariant under M_R. Similarly $(DI)_o^-$ is invariant under T_D which is Zarisky dense in $A_h(T_D) \supset T_R$ and hence $(DI)_o^-$ is invariant under T_R. This shows that $(DI)_o^-$ is normal in R_S.

Now let R/D be a presentation of a compact solvmanifold and let $p(\xi)$ be a G-induced flow in general position. We will seek to find an equivalent G-induced flow $p^*(\xi)$ on R^*/D^*, where D/D^* is finite such that R^* satisfies the hypothesis of Theorem 6. (This is essentially the method of modification as introduced in [3].)

As usual let $R_S = M_R \rtimes T_R$ and let $D_S = M_D \rtimes T_D$, where $M_D \subset M_R$ and $T_D \subset T_R$. Since T_R commutes with T_D, T_R commutes with $A_h(T_D)$, the algebraic hull of T_D. Now $D/N \cap D = \mathbb{Z}^s$ for some s where N is the nil-radical of R. Hence we have a surjection

$$\alpha: \mathbb{Z}^s \longrightarrow T_D$$

Let D^* be the subgroup of D of finite index such that

$$\alpha(D^*/D \cap N) = T_D \cap A_h(T_D)_o$$

Then $D^*/D \cap N$ is also isomorphic to $\mathbb{Z}^s$ and we may form $(D^*/D \cap N) \otimes \mathbb{R}$ and extend α to a homomorphism

$$\alpha^* : (D^*/D \cap N) \otimes \mathbb{R} \longrightarrow A_h(T_D)_o$$

Let $((D^*/D \cap N) \otimes \mathbb{R})\alpha^* \ltimes M_R$ denote the semi-direct product of $(D^*/D \cap N) \otimes \mathbb{R}$ and M_R where $(D^*/D \cap N) \otimes \mathbb{R}$ acts on M_R as its image under α^*. It is easy to verify that in $((D^*/D \cap N) \otimes \mathbb{R})\alpha^* \ltimes M_R$ we may find a connected, simply connected group R^* with the following properties:

(1) $R^* \supset D^*$ and R^*/D^* is compact

(2) if $R_S^\sim$ denotes the universal covering group of R_S then

$$R_S^\sim = ((D^*/D \cap N) \otimes \mathbb{R})\ \alpha^* \ltimes M_R$$

(3) There is a compact abelian group of automorphisms C such that $R \subset C \ltimes R^*$

(4) Let q be the projection of $C \ltimes R^*$ onto R^*, then q induces a homeomorphism

$$q^*: R/D^* \longrightarrow R^*/D^*$$

Now Lemma 1 of [3] and some minor verifications imply Theorem 7 below.

Theorem 7. Let R/D^* be a presentation of a compact solvmanifold and let $p(\xi)$ be a G-induced flow in general in R/D^*. Assume further that D^* satisfies the condition that $A_h(T_D)_o$ is connected. Then there exists $R^* \supset D^*$ such that R^*/D^* is compact and $T_{R^*} \subset A_h(T_D)_o$. Further there exists $p^*(\xi)$ a one parameter subgroup of R^* in general position in R^*/D^* such that

1) the flow induced by $p^*(\xi)$ on R^*/D^* is equivalent to the flow induced by $p(\xi)$ on R/D^*.

2) the unipotent part of $p^*(\xi)$ is the same as that of $p(\xi)$

3) $p^*(\xi)$ and $p(\xi)$ have the same constable normal subgroup in M_R.

Theorem A. Let $p(\xi)$ be a G-induced flow on the presentation R/D of a compact solvmanifold. The following are necessary and sufficient conditions for $p(\xi)$ to act ergodically on R/D:

1) $p(\xi)$ is in general position

2) Let $\rho: M_R \longrightarrow M$ be the natural homomorphism with kernel D_o and let $B = (\rho(U)^-)_o$, where U is the unstable subgroup of R relative to $p(\xi)$. If T^* is the semi-simple part of D^* acting on

M_R/D_o then B is a normal subgroup of $M \rtimes T^*$ and

$$(M \rtimes T^*)/B = M/B \oplus T^*$$

3) The unipotent part of $p(\xi)$, $m(\xi)$, acts ergodically on $M/B\Gamma^*$, where $\Gamma^* = D^*/D_o$

<u>Proof</u>: Since statements 1, 2 and 3 above involve only M_R, M, D, $m(\xi)$ and $p(\xi)$ we may replace R/D with R^*/D^* and $p(\xi)$ by $p^*(\xi)$ in the hypothesis of Theorem A. We may now apply theorem 6 to conclude that $((D^*U)^-)_o$ is normal in R^*. Since $R^*((D^*U)^-)_o$ is class R, we have that two above is satisfied. We may now apply the Corollary to Theorem 2 to prove our theorem.

It is important to note that condition 2 is really implies by condition 1.

Since $M/B\Gamma^*$ is a nilmanifold we may form an abelized version of Theorem A.

<u>Theorem A'</u>. Using the above notation $p(\xi)$ is ergodic on R/D if and only if $p(\xi)$ is in general position and $m(\xi)$ acts ergodically on $(M/B)/([M/B, M/B]\Gamma^*B/B)$, where $\Gamma^* = D^*/D_o$.

Thus to make theorem A really useful all that is necessary is to give a simple description of the manifold occuring in Theorem A'.

Before doing this let us recall some basic facts about the abelian theory. Let V^n be an n dimension vector space over the reals and let $e_1, \dots, e_n$ be a basis of V^n and let $V^n_{\mathbb{Q}}$ be the rational points of V^n relative to this basis; i.e. $V^n_Q = \{\sum_{i=1}^{n} q_i e_i \mid q_i \in \mathbb{Q}\}$. We will call $V^n_{\mathbb{Q}}$ a rational form of V^n. Clearly V^n has lots of distinct rational forms and $V^n_{\mathbb{Q}}$ is an n-dimensional vector space over $\mathbb{Q}$. Let L be a lattice in $V^n_{\mathbb{Q}}$. Now let A be a linear subspace of V^n, then $((LA)^-)_o$ is the real points of the smallest rational subspace whose real points contain A.

Now let $p(\xi)$ be a line in V^n. Then $p(\xi)$ acts ergodically on V^n/L if and only if $((p(\xi)L)^-)_0 = V^n$.

Since M/B is a homomorphic image of M we may describe $(M/B)/[M/B, M/B]$ by describing the subspace K of $M/[M,M]$ that is the kernel of $M/[M,M] \longrightarrow (M/B)/[M/B, M/B]$.

Theorem B. Let R/D be a presentation of a compact solvmanifold with fundamental group Γ and let $p(\xi)$ be a G-induced flow in general position with unipotent part $m(\xi)$. Let $M \rtimes T$ be the semi-simple splitting of Γ with connected nil-shadow and let $V = M/[M,M]$ with its induced rational structure. Then $p(\xi)$ acts ergodically in R/D if and only if the image of $m(\xi)$ on V satisfies the following condition:

Let W be the subspace of V consisting of eigen-vectors of some element of T whose eigen-values has non-trivial real part. Let $B_{\mathbb{Q}}^*$ be the smallest rational subspace whose real points B^* contain W. Then the image of $m(\xi)$ in V/B^* is in general position; i.e., the smallest rational subspace whose real points contain the image of $m(\xi)$ is the whole space V/B^*.

Proof: All we have to verify is that condition 3 is implied by the condition of Theorem B. Since $D \cap [M_R, M_R]$ is cocompact we may reduce our Theorem to the special case where $[M_R, M_R]$ is trivial. But, since the homomorphic image of a class R group is class R, it follows that in our special case the B of Theorem A is a rational subspace of V whose real points contain W. It is also clear that $B \subset B^*$, since $(\rho(U)^-)_0$ is the smallest rational subspace that contains a subset of W. Hence $B = B^*$ and we have proven our theorem.

It is not difficult from this result to prove the following:

Corollary: Let R/D be the presentation of a compact solvmanifold and let R be class R. Then if R/D is not a nilmanifold it possess no G-induced flow.

References

[1]. L. Auslander et al., Flows on homogeneous spaces, Annals of Math. Study 53 (1963)

[2]. ________ and L. Green, G-Induced Flows, Amer. J. of Math. 88 (1966) p. 43-60.

[3]. ________, Modifications of solvmanifolds and G-induced flows, Amer. J. of Math. 88 (1966) p. 615-625.

[4]. ________, An exposition of the algebraic theory of solvmanifolds, to appear Bull. A. M. S.

PRODUCTS OF SEMI-DYNAMICAL SYSTEMS

Prem N. Bajaj

Wichita State University

INTRODUCTION

Semi-dynamical systems (s.d.s.) are continuous flows defined only in the 'future'. Functional differential equations for which existence and uniqueness conditions hold [6] provide natural examples. Many new and interesting notions (e.g., a start point) arise in s.d.s. Moreover, s.d.s. generalize a considerable portion of the theory of dynamical systems. The idea of a start point dates back to, at least, 1953, when [5] appeared. For a family of s.d.s., the product is defined in a natural way.

This paper is divided into three sections. In the first section basic concepts are introduced. The second section deals with limit sets and the last section with start points. Start points in a product s.d.s. can arise without any of the factor systems having any. Notion of improper start point is introduced and necessary and sufficient conditions obtained for their existence. In the presence of an improper start point, the set of start points is dense everywhere, and (path) connectedness of any of the sets of start points or improper start points or the product space are equivalent.

I BASIC NOTIONS

1.1. Definitions. Let X be a Hausdorff topological space and R^+ the set of non-negative reals with usual topology. Then a continuous map π from $X \times R^+$ into X is said to define a semi-dynamical system (s.d.s.) (X, π) if $\pi(x, 0) = x$ and $\pi(\pi(x, t), s) = \pi(x, t+s)$ for each x in X, and t,s in R^+. For brevity (as in [1], [3]), we denote $\pi(x, t)$ by xt, the set $\{xt: x \in M \subset X, t \in K \subset R^+\}$ by MK, etc.

Define a map $E: X \to R^+ \cup \{+\infty\}$ by $E(x) = \{t \geq 0: yt = x$ for some y in $X\}$. $E(x)$ is called the escape time of x. Throughout E will denote the map defined above. A point x is said to be a start point if its escape time vanishes. Positive trajectory $\gamma^+(x)$ of any point x is the set xR^+. If $\gamma^+(x) = \{x\}$, x is (positively) critical.

A semi-dynamical system (X, π) is said to have unicity if $xt = yt$ implies $x = y$ for all x,y in X and $t \in R^+$.

The (positive) limit set $\Lambda(x)$ of any point x in X is the set $\{y \in X:$ there exists a net t_i in R^+, $t_i \to +\infty$ such that $xt_i \to y\}$.

1.2. Proposition. Let (X_α, π_α), $\alpha \in I$ be a family of semi-dynamical systems. Let $X = \prod X_\alpha$ be the product space. Let $x \in X$, $x = \{x_\alpha\}$. Define a map π from $X \times R^+$ into X by $\pi(x, t) = \{x_\alpha t\}$. Then (X, π) is a semi-dynamical system.

The s.d.s. (X, π) obtained above is called the direct product (or simply product) of the family (X_α, π_α), $\alpha \in I$ of s.d.s. Clearly (X, π) has unicity if and only if each (X_α, π_α), $\alpha \in I$ has unicity.

II LIMIT SETS

2.1. Proposition. Let (X, π) be a semi-dynamical system. Let X be rim compact [7, p. 111]. Let $x \in X$. If $\Lambda(x)$ is non-empty and compact, then $Cl(xR^+)$ is compact.

Proof. Notice that given a neighborhood U of $\Lambda(x)$, there exists $T > 0$ such that $xt \in U$ for every $t \geqslant T$, etc., in particular, if t_i is a net in R^+, $t_i \to +\infty$, then $xt_i \to \Lambda(x)$.

2.2. Proposition. Let (X_α, π_α), $\alpha \in I$ be a family of s.d.s. and (X, π) the product s.d.s. Let $x \in X$, $x = \{x_\alpha\}$. Then $\Lambda(x) \subset \prod \Lambda_\alpha(x_\alpha)$ where $\Lambda_\alpha(x_\alpha)$ is the positive limit set of x_α in the s.d.s. (X_α, π_α)

Similar relations hold in a product s.d.s. for positive prolongations and their limit sets defined in the natural way [1], [3], [8].

In general, equality does not hold in the above. Indeed $\Lambda(x)$ may be empty even if each of $\Lambda_\alpha(x_\alpha)$, $\alpha \in I$ is non-empty. We state two theorems; one in which $\Lambda(x)$ is non-empty and the second indicating a case of equality.

2.3. Theorem. Let (X_j, π_j), $j = 1,2, \ldots, n$ be s.d.s. and (X, π) their product, $n > 1$. Let the spaces $X_1, X_2, \ldots, X_{n-1}$ be rim-compact. Let $x = (x_1, \ldots x_n) \in X$. If $\Lambda_j(x_j)$, $j = 1,2, \ldots n$ is non-empty and compact, then $\Lambda(x)$ is non-empty and compact.

2.4. Theorem. Let (X_1, π_1), (X_2, π_2) be s.d.s. and (X, π) their product. Let X_1 be rim-compact. Let $x = (x_1, x_2) \in X$. Let $\Lambda_1(x_1) =$

$\{a\}$ be a singleton. If $\Lambda_2(x_2)$ is non-empty, then $\Lambda(x) = \Lambda_1(x_1) \times \Lambda_2(x_2)$.

III START POINTS

3.1. Theorem. Let (X_α, π_α), $\alpha \in I$ be a family of s.d.s. and (X,π) the product s.d.s. Let S_α denote the set of start points in (X_α, π_α) for any α in I, and S the corresponding set in (X, π). Then $S \supset U_\alpha(S_\alpha \times \prod_{\beta \neq \alpha} X_\beta)$. Moreover, equality holds if I is finite.

Proof. Notice that if $x \in X$, $x = \{x_\alpha\}$, and x_α is a start point for some α, then x is a start point, etc. If I is finite, and $x \in X$ is a start point then x_α is a start point for some α; indeed, otherwise, $E(x) = \inf \{E(x_\alpha): \alpha \in I\} > 0$ which contradicts that x is a start point.

3.2. Remark. If I is infinite, equality does not hold, in general, in the above theorem. Start points in factor s.d.s. give rise to those in a product s.d.s. However, start points in a product s.d.s. can arise without the factor s.d.s. having any. We have the following:

3.3. Definition. Let (X_α, π_α), $\alpha \in I$ and (X, π) be as above. Let $x \in X$, $x = \{x_\alpha\}$ be a start point. Then, relative to the factorization $\prod X_\alpha$ of X, start point x is said to be proper if x_α is a start point for some α in I; otherwise, call the start point improper.

The following theorem indicates how the improper start points arise.

3.4. Theorem. (Preserving the notation). In the product s.d.s. (X,π), there exists an improper start point if and only if for infinitely many α in I the s.d.s. (X_α, π_α) contains a point x_α with finite non-zero escape time.

Proof. If $x \in X$, $x = \{x_\alpha\}$ is an improper start point, then $E(x_\alpha) > 0$ for every α. Clearly $\{\alpha\colon E(x_\alpha) < 1\}$ is infinite. Indeed, otherwise, $E(x) = \inf\{E(x_\alpha)\colon \alpha \in I\} > 0$; contradiction as x is a start point.

Conversely, for each positive integer n, pick an α_n in I such that $0 < E(x_{\alpha_n}) < +\infty$ for some x_{α_n} in X_{α_n}. Choose y_{α_n} in X_{α_n} such that $0 < E(y_{\alpha_n}) \leqslant \frac{1}{n}$. Let $T > 0$ be fixed. Then $z \in X$, $z = \{z_\alpha\}$ such that $z_{\alpha_n} = y_{\alpha_n}$ for every n, and $z_\alpha = x_\alpha T$ otherwise, where $x_\alpha \in X_\alpha$ is arbitrary, is an improper start point.

3.5. Theorem. [2, p. 175]. Let (X_α, π_α), $\alpha \in I$ and (X, π) be as above. The set of start points is dense in X if and only if at least one of the following conditions hold:

(a) There exists an improper start point.

(b) The set of start points is dense in X_β for some β in I.

(c) Infinitely many factor s.d.s. contain start points.

In a topological space, closure of a path connected set is not, in general, path connected. Indeed in a topological space Y, if $K \subset H \subset \text{Cl}\, K \subset Y$, both K, $\text{Cl}\, K$ are path connected, even then H is not necessarily so. In s.d.s., in the presence of an improper start point, we have the following:

3.6. Theorem. Let (X_α, Π_α), $\alpha \in I$ and (X, Π) be as above. Let there exist an improper start point. The following are equivalent:

(a) The set of start points in (X, Π) is (path) connected.

(b) X is (path) connected.

(c) The set of improper start points is (path) connected.

Proof. We outline only one case.

Let the set S of start points in (X, Π) be path connected. Let $x_\beta, x'_\beta \in X_\beta$ be arbitrary. Let $z = \{z_\alpha\}$ be an improper start point. Pick y, y' in X, $y = \{y_\alpha\}$, $y' = \{y'_\alpha\}$ such that $y_\beta = x_\beta$, $y'_\beta = x'_\beta$ and $y_\alpha = y'_\alpha = z_\alpha$ otherwise. Let $f: J \to S$ be a path joining the start points y and y', (J is the closed unit interval). Then $\mathrm{proj}_\beta \circ f: J \to X_\beta$ is a path joining x_β and x'_β. Thus X_β is path connected. Since β is arbitrary, it follows that X is path connected.

REFERENCES

[1] J. Auslander and P. Seibert, Prolongations and Stability in Dynamical Systems, Ann. Inst. Fourier, Grenoble, 14 (1964), pp. 237-268.

[2] Prem N. Bajaj, Start Points in Semi-dynamical Systems, Funkcialaj Ekvacioj, 13 (1971), pp. 171-177.

[3] N. P. Bhatia and G. P. Szegö, Dynamical Systems, Stability Theory and Applications, Springer-Verlag, New York, 1967.

[4] W. Gottschalk and G. Hedlünd, Topological Dynamics, Colloquium Publications, Volume XXXVI, American Mathematical Society, 1955.

[5] I. Flügge Lots, Discontinuous Automatic Control, Princeton University Press, 1953.

[6] J. Hale, Sufficient Conditions for Stability and Instability of Autonomous Functional Differential Equations, J. Differential Equations, 1 (1965), pp. 452-482.

[7] J. R. Isbell, Uniform Spaces, Math. Surveys, No. 12, American Math. Society, 1964.

[8] T. Ura, Sur Le Courant Extérieur à une région invariante, Funk. Ekv., 2 (1959), pp. 143-200.

[9] S. Willard, General Topology, Addison-Wesley, Reading, Massachusetts, 1970.

UNIQUENESS OF FLOW SOLUTIONS OF DIFFERENTIAL EQUATIONS

ANATOLE BECK

UNIVERSITY OF WISCONSIN, MADISON

1. INTRODUCTION. The foundation stones of the theory of Differential Equations are the theorems of Cauchy and Picard assuring the existence and uniqueness of solutions of certain differential equations. We state them here:

1.1 THEOREM (Cauchy-Peano): Let f be a continuous function from a domain $D \subset \mathbb{R} \times \mathbb{R}^n$ into $\mathbb{R}^n$, and let $(t_o, x_o) \in D$.

Then the differential equation $\dot{x} = f(t,x)$ has a solution g in $\mathbb{R}^n$ satisfying $g(t_o) = x_o$.

1.2 THEOREM (Picard-Lindelof): Let f be a continuous function from a domain $D \subset \mathbb{R} \times \mathbb{R}^n$ into $\mathbb{R}^n$, and let $(t_o, x_o) \in D$. Assume there exists a number k so that for all $t \in \mathbb{R}$, and all $x_1, x_2 \in \mathbb{R}^n$ satisfying $(t,x_1), (t,x_2) \in D$, we have

$$||f(t,x_1) - f(t,x_2)|| < k \, ||x_1 - x_2|| \, .$$

Then the differential equation $\dot{x} = f(t,x)$ has a unique solution g in $\mathbb{R}^n$ satisfying $g(t_o) = x_o$.

That the condition of continuity is not sufficient for uniqueness is seen easily in the following example:

1.3 EXAMPLE: Let the function f from $\mathbb{R} \times \mathbb{R}$ into $\mathbb{R}$ be defined by

$$f(t,x) = \begin{cases} 2t & \text{if } x > t^2 \\ \frac{2x}{t} & \text{if } |x| \leq t^2 \\ -2t & \text{if } x < -t^2 \end{cases}$$

We see easily that the solutions g satisfying $g(0) = 0$ include all the functions at^2 for $|a| \leq 1$.

The "ultimate example" in this regard is to be found in Hartman [1] pp. 18ff. There, we find a continuous differential equation in the plane such that for every point in the plane, and every neighborhood of that point, there exist $\aleph\ (= 2^{\aleph_0})$ different solutions through the point no two of which agree in the indicated neighborhood.

An interesting question is whether the addition of qualitative conditions will enable the above result to hold with continuity only. The most obvious result is one of "autonomy", i.e. the function $f(t,x)$ is actually a function of x alone. This is obviously wrong, since we can make any differential equation into an autonomous one by running up the dimension by one. Another possible requirement is that the solution curves be very smooth; we shall hear more of this later. But the most plausible requirement is that the solution of the differential equation be not merely a collection of curves (or of functions) but actually a whole family of such functions constituting a flow or an action.

2. FLOWS We define as an <u>algebraic flow</u> a mapping φ from $\mathbb{R} \times X$ into X (where X is any given set) which satisfies the group property:

$$\varphi(t, \varphi(s,x)) = \varphi(t+s, x) ,$$

$\forall t,s \in \mathbb{R}$, $\forall x \in X$. If, in addition, we put another condition on φ, we obtain a special category of flow. In this case, we take X to be one of the spaces $\mathbb{R}^n$ and require of φ that at every $x \in \mathbb{R}^n$, the limit

$$\dot{\varphi}(x) = \lim_{t \to 0} \frac{1}{t} (\varphi(t,x) - x) \quad \text{exist.}$$

Then we call φ a <u>velocity flow</u> in $\mathbb{R}^n$. If $\dot{\varphi}$ is a continuous function from $\mathbb{R}^n$ into itself, then we call φ a <u>continuous velocity</u>

flow. If $\varphi(x) \neq 0, \forall x \in \mathbb{R}^n$, then φ is called non-singular.

Working from this definition, we could ask the question: Do there exist two different non-singular continuous velocity flows whose velocity functions are identical? The motivation for asking the question in this form is, in part, the following example:

2.1 EXAMPLE: Let a flow φ be defined in $\mathbb{R}^2$ so that the velocity $\dot{\varphi}$ at that point (x,y) is given by $(1, 3y^{2/3})$. Then φ must be given by $\varphi(t,(x,0)) = (x+t, t^3)$ or $\varphi(t,(x,y)) = (x+t, (t+y^{1/3})^3)$. The orbits of this flow are the curves $y = (x+C)^3$, and these curves are thus solution curves of the differential equation as well. However, there are other solution curves. For any choice of numbers a,b satisfying $-\infty \leq a \leq b \leq +\infty$, we can define $g_{a,b}(x)$ to be equal to $(x-a)^3$ if $x \leq a$, 0 if $a \leq x \leq b$, and $(x-b)^3$ is $x \geq b$. Then each of the curves $y = g_{a,b}(x)$ is a solution curve of the differential equation $y' = 3y^{2/3}$.

We note, however, that $y = g_{a,b}(x)$ cannot be an orbit of any flow satisfying the indicated differential equation, unless $a = b$. To show this, let c be chosen between a and b. Then we observe that for any flow Ψ satisfying the differential equation, we would have $\Psi(c-b, (b,(b-c)^3)) = (c,0)$, while $\Psi(c-b,(b,0)) = (c,0)$. Thus, the group property gives us $\Psi(b-c,(c,0))$ as both $(b,0)$ and $(b,(b-c)^3)$, which is a contradiction. It follows that there is no such flow Ψ. It follows from this that although each point in the plane has $\aleph$ different solutions of the differential equation passing through it, there is only one flow which satisfies the velocity condition.

2.2 QUESTION: Is it possible to find two different non-singular velocity flows in $\mathbb{R}^n$ with the same velocity function?

2.3 ANSWER Yes, and in a very strong sense, if $n \geq 2$.

3. THE HARE AND THE TORTOISE We will exhibit a class of flows in the plane, $\mathbb{R}^2$, with $\aleph$ members, all of which have the same velocity function. Since there are only $\aleph$ flows possible in the plane, we are at least maximizing as to number.

3.1 EXAMPLE Let K be an Cantor set in $[0,1]$, and let g be any member of $C^{\infty}(\mathbb{R})$ satisfying $0 \leq g(x) < 1$, $\forall\, x \in \mathbb{R}$, and $g(x) = 0$ iff $x \in K$. Let $M_0(K)$ (denoted as M_0) be the set of atom-free, positive, finite measures on K. Let g be integrated to yield the function f :

$$f(x) = \int_0^x g(t)\, dt\ , \quad \forall\, x \in \mathbb{R}\ .$$

For each measure $m \in M_0$, let f_m be the function defined by $f_m(x+m(K \cap [0,x])) = f(x)$. Then for each m, we define φ_m by $\varphi_m(t,(x,0)) = (x+t,\ f_m(t))$, or $\varphi_m(t,(x,y)) = (x+t, f_m(t+f_m^{-1}(y)))$, $\forall\, t \in \mathbb{R}$, $\forall\, (x,y) \in \mathbb{R}^2$. The orbits of the flow φ_m are precisely the curves $y = f_m(x+C)$, and we see at once that the velocity $\dot{\varphi}_m$ at the point (x,y) is exactly $(1,\ f_m'(f_m^{-1}(y)))$. Our next step is the evaluation of this last expression.

For each $t \notin K$, we have $s \notin K$ for all s in some neighborhood N of t. It follows that for all $s \in N$, $f_m(s + m(K \cap [0,s])) = f_m(s + m(K \cap [0,t])) = f(s)$. Thus, $f_m'(t + m(K \cap [0,t])) = f'(t)$, $\forall\, t \notin K$. If $f(t) = y$, then $f^{-1}(y) = t$ and $f_m^{-1}(y) = t + m(K \cap [0,t])$, so that $f_m'(f_m^{-1}(y)) = f_m'(t + m(K \cap [0,t])) = f'(t) = f'(f^{-1}(y))$ whenever $y = f(t)$ and $t \notin K$.

For each $t \in K$, we have $f'(t) = g(t) = 0$. On the other hand, to reckon $f_m'(t + m(K \cap [0,t]))$, we note that for each $t_1 < t_2$, we have

$$\begin{aligned}(t_2 + m(K \cap [0,t_2])) - (t_1 + m(K \cap [0,t_1])) &= t_2 - t_1 + m(K \cap [t_1,t_2]) \\ &\geq t_2 - t_1 .\end{aligned}$$

Thus, $\dfrac{y_2 - y_1}{f_m^{-1}(y_2) - f_m^{-1}(y_1)} \le \dfrac{y_2 - y_1}{f^{-1}(y_2) - f^{-1}(y_1)}, \quad \forall\, y_1 < y_2 .$

If $y = f(t)$, then

$$0 \le \limsup_{y_1 \to y} \frac{y - y_1}{f_m^{-1}(y) - f_m^{-1}(y_1)} \le \lim_{y_1 \to y} \frac{y-y_1}{f^{-1}(y)-f^{-1}(y_1)}$$

$$= f'(t)$$

$$= 0 ,$$

so that $f_m'(t+m(K\cap[0,t]))$ exists and equals 0. It follows that when $y = f(t)$, with $t \in K$, we again have $f_m'(f_m^{-1}(y)) = f'(f^{-1}(y))$.

Thus, in either case, we have this last equality, i.e. it holds for all $y \in \mathbb{R}$. It follows that $\dot{\varphi} = \dot{\varphi}_m$ everywhere in $\mathbb{R}^2$ for any $m \in M_o$.

3.2 COROLLARY: It is possible to construct two continuous velocity flows in $\mathbb{R}$ which have the same non-negative velocity at each point such that neither has a fixed point, and such that the time taken by the second to cover the interval [0,1] exceeds the time taken by the first by any desired quality.

PROOF: Let the flow Ψ be defined in $\mathbb{R}$ by $\Psi(t,x) = f(t+f^{-1}(x))$. Choose any positive number y and select a measure $m \in M_o$ such that $m(K) = y$. Define the flow Ψ_m by $\Psi_m(t,x) = f_m(t + f_m^{-1}(x))$. Then by the previous example, we see that $\dot{\Psi}(x) = \dot{\Psi}_m(x), \ \forall\, x \in \mathbb{R}$, while $\Psi(f^{-1}(1), 0)) = 1 = \Psi_m(y + f^{-1}(1), 0))$. QED

3.3 COROLLARY. (The Hare and the Tortoise): There exist two continuous velocity flows Ψ^- and Ψ^+ in $\mathbb{R}$ which have no fixed points, and a positive number t_o such that

i) $\dot{\Psi}^-(x) = 100\,\dot{\Psi}^+(x), \quad \forall\, x \in \mathbb{R}$,

ii) $\Psi^+(t_o,0) = 1$.

iii) $\Psi^-(t_o,1) < 1$.

PROOF: Returning to the notation of Example 3.1 and Corollary 3.2, let y be chosen greater than $99f^{-1}(1)$ and let m be chosen so that $m(K) = y$. Let f^+ be defined by $f^+(100x) = f(x)$. Define Ψ as in Corollary 3.2, and Ψ^- to be the flow Ψ_m defined there. Define Ψ^+ by $\Psi^+(t,x) = f^+(t + f^{+^{-1}}(x))$. Then clearly, $\Psi^-(x) = \Psi(x)$, $\forall x \in \mathbb{R}$, and also $\dot{\Psi}(x) = 100\,\dot{\Psi}^+(x)$. If we set $t_o = 100\, f^{-1}(1)$, then $\Psi^+(t_o,0) = 1$, while $\Psi^-(t_o,0) < 1$, since $t_o < y + f^{-1}(1)$. QED

3.4 EXAMPLE: We will now exhibit another example, similar to Example 3.1, but with an added wrinkle. We will construct a flow φ_m with the properties that

i) $\dot{\varphi}_m(x,y) = \dot{\varphi}(x,y)$, $\forall (x,y) \in \mathbb{R}^2$.

ii) $\exists\, y_o$ such that the line $y = y_o$ is an orbit of φ_m.

Let t_o be a point of K which is not 0 or 1 nor an endpoint of any of the intervals complimentary to K. Define the sequences $\{t_n^-\}$ and $\{t_n^+\}$ in the following way:

i) $t^-_1 = 0$, $t^+_1 = 1$.

ii) $t^-_n \uparrow t_o$, $t^+_n \downarrow t_o$.

We now define a measure m on K so that

$$m(K \cap [t^-_n, t^-_{n+1}]) = m(K \cap [t^+_n, t^+_{n+1}]) = 1, \forall n,$$

and m is atom-free. Let f_m^- and f_m^+ be defined by

$$f_m^-(t + m(K\cap[0,t])) = f(t), \quad \forall\, t < f^{-1}(y_o),$$

$$f_m^+(t - m(K\cap[t,1])) = f(t), \quad \forall\, t > f^{-1}(y_o),$$

note that both f_m^- and f_m^+ are defined for all real values of the argument. Define f_m^{-1} to be the inverse of f_m^- or of f_m^+ according as the value of the argument is less or greater than y_o.

We define the flow φ_m in $\mathbb{R}^2$ by

$$\varphi_m(t,(x,y)) = \begin{cases} (t+x,\ f_m^-(t + f_m^{-1}(y)\ )) & \text{if } y < y_o \\ (t+x,\ y_o) & \text{if } y = y_o \\ (t+x,\ f_m^+(t + f_m^{-1}(y)\ )) & \text{if } y > y_o \end{cases}$$

It is now clear that φ_m has the desired properties.

3.5 DEFINITION: We define as $M(K)$ (denoted as M) the set of all atom-free measures on K which are σ-finite and which have the additional property that for each point $t \in K$, either there exist intervals on both sides of t with t as endpoint and finite measure, or else every interval on either side of t which has t for an endpoint has infinite measure.

For every measure $m \in M$, we see at once that the set of all t such that every interval with t as endpoint has infinite measure is a closed set. Thus, its compliment is open, and is thus the union of countably many (maybe finitely many) open intervals. For each such interval U, we define the function f_m^U as follows: Let t_U be the midpoint of U, and $y_U = f(t_U)$. For $t \in U$, we define $f_m^U(t + m(K \cap [t_U, t])) = f(t)$ if $t > t_U$, and $f_m^U(t - m(K \cap [t, t_U])) = f(t)$ if $t < t_U$. Then the range of f_m^U will be exactly $f(U)$.

We define f_m^{-1} by defining its values in each such interval $f(U)$: f_m^{-1} in $f(U)$ is defined to be the inverse of f_m^U. Now we define the flow φ_m:

If y belongs to one of the intervals U, then $\varphi_m(t,(x,y)) = (x+t,\ f_m^U(t + f_m^{-1}(y)\ ))$. If not, then $\varphi_m(t,(x,y)) = (x+t,\ y)$.

3.6 THEOREM: For each $m \in M(K)$, φ_m is a flow in $\mathbb{R}^2$, and $\dot{\varphi}_m(x,y) = \dot{\varphi}(x,y)$, $\forall\ (x,y) \in \mathbb{R}^2$.

PROOF: Clear.

3.7 DEFINITION: For every closed, nowhere dense set K, we can define $M_0(K)$ and $M(K)$ as we did for the Cantor set. We can also define a function $g \in C^\infty(\mathbb{R})$ which vanishes precisely on K and takes values between 0 and 1. Integrating g to obtain f, we can define the flow φ_m as before, for each $m \in M(K)$. Clearly, the set $M(K)$ has non-zero elements iff K is uncountable.

4. "HARTMANIZING" SECTION 3: As we have seen, there are $\aleph$ different flow solutions for the vector field $\dot{\varphi}(x,y)$. However, if $y \notin f(K)$, then there is a neighborhood of (x,y) in which the flow is unique, i.e. in which all the flow solutions agree. Our next step is to find a way of killing this uniqueness.

4.1 EXAMPLE: We will now construct a non-vanishing, continuous vector field in $\mathbb{R}^2$ with the property that for any point $(x,y) \in \mathbb{R}^2$, and for any neighborhood N of (x,y), there exist $\aleph$ non-singular continuous velocity flows in R^2 so that no two of them have orbits through (x,y) which agree in N.

We start by rebuilding Example 3.1 to meet our new needs. In place of the Cantor set, we define K to be the union of infinitely many Cantor sets, one in each of the intervals $[n, n+1]$, $-\infty < n < +\infty$. We define g in the same way, requiring that $g(x) = g(x+1)$, $\forall x \in \mathbb{R}$, and obtain f, as before, by integrating g. We then obtain the continuous non-singular vector field $\dot{\varphi}(x,y)$ in the same way as before, except that now, we normalize all the vectors to be unit vectors, and call the field $V_1(x,y)$. If $y \in f(K)$, then for any neighborhood of (x,y), there are $\aleph$ functions f_m so that no two agree in that neighborhood, and so the field $V_1(x,y)$ satisfies our requirements at those points. Furthermore, if (x,y) does not satisfy $y \in f(K)$ but there is, in the given neighborhood, a point of its orbit which does satisfy that condition, then again we can satisfy the conclusion of the example. We will examine the action of a solution flow φ_1 of V_1 in the strip $R \times U$, where $U = (u^-, u^+)$ is a component of $\mathbb{R}\setminus f(K)$. For any point $(x,y) \in \mathbb{R} \times U$, the orbit of any solution of the vector field is unique within the strip, and "goes to pieces" as soon as it reaches the boundary. We will change the field in this strip by changing the directions of the unit vectors as little as we wish, and making the maximal regions of unicity much much smaller. Note that the largest disc which can be included in any "region of uniqueness" has diameter less than 1/3, which is greater than the width of any of the strips $\mathbb{R} \times U$. Note also that all the vectors have for their slopes certain values of $g(x)$, and that these are all bounded by 1, so that the angle of inclination of all these vectors is less than $\frac{\pi}{4}$.

Next we will modify the tangent field $V_1(x,y)$. For each strip $\mathbb{R} \times U$, interchange the roles of x and y. This will give us the strip $U \times \mathbb{R}$, and in that strip, the vector field becomes the unit tangents to the curves $y = f^{-1}(x) + C$. To see how to modify the vector field, the first step is to consider the unit tangents to the curves $u = f(\frac{x}{u}) + C$ in the same strip, where $u = u^+ - u^-$. Since the width of the strip is u, we see that each of these curves increases by the amount $f(1)$ in the strip U, and thus passes through a whole copy of the set $f(K)$, mod $f(1)$. We now multiply the strip in the y direction by a positive number $b_1 < 1$. The result is that the curves become the curves $y = b_1\ f(\frac{x}{u}) + C$, so that at every point, the slope of the tangent is no more than b/u. Thus, if $b < \frac{1}{2}u$, we know that the slope of each such tangent is less than 1/2. Now consider at each point the unit vector whose slope is the sum of these two slopes, i.e.

$$\frac{1}{g(f^{-1}(x))} + \frac{b_1}{u}\ f'(f^{-1}(\tfrac{y}{b}))\ .$$

We see that this field of vectors differs in direction from the inverted field $V_1(x,y)$ by an angular amount which is not more than 1/2 rad. Now, if we again reverse the roles of x and y, then we have created a new field of vectors, a continuous field of unit vectors which differ from the field V_1 by an angle of less than 1/2 rad. If we consider all the flows which have these vectors for their velocities, we see that not only are the lines $y \in f(K)$ choice points, but also the images of these lines in $\mathbb{R} \times U$, i.e. the lines in U satisfying

$$\frac{x}{b_1} - f^{-1}\left(\frac{y}{u}\right) \in f(K)$$

We see at once that the largest disc of uniqueness has diameter less than $\frac{1}{3} b_1$, which we can make as small as we like.

What we have done for the strip U we can do for each strip in the set $\{(x,y) \mid y \in f(K)\}$, where $b_1(U)$ varies with each choice of U. Note that the longest orbit arc penetrating a region of uniqueness under the vector field V_1 is no longer than $2/3$ in length, and that by choosing b_1 small enough for each U, we can assure that the longest orbit arc penetrating a region of uniqueness of the new field is less than half that long. (In fact, we can get as close as we like to making it no more than $1/3$ that long.)

The field we obtain in this way differs from the field V_1 by no more than $\frac{1}{2}$ rad at each point. In order to assure the convergence of the sequence of vector fields we shall construct, we normalize the new field to give us a field of unit vectors, which we call $V_2(x,y)$.

Now comes the hard part. To iterate this process, we will use a "backward" method in a sense we shall see shortly. We want to change the vector field only at those points which are in the region of uniqueness, and decimate this region of uniqueness with new "choice points" by an arbitrarily small change in the field V_1 so that all the previous choice points retain their character. For each of the strips $\mathbb{R} \times U$, we will modify the field not by the compressed part of the field used above, but by a compressed part of a field which has already been modified.

For each pair of strips U_1 and U_2 which are components of $\mathbb{R}\setminus f(K)$, let us define $b_2(U_1,U_2)$ (which we shall now call merely b_2), and carry out the modifications described above for the strip U_2.

Reserving comment on the size of b_2, we now use the modified vector field to modify the portion of V_1 lying in U_1. That is, we modify using a modified field instead of the unmodified field we had before, but we use the factor $b_1(U_1)$ which we selected before. The result is a new field, which is just like the field V_1 except in the part of the plane previously modified, i.e. in the portion corresponding to U_2 two modifications back. We now see that as b_2 converges to 0, the modification converges to 0 uniformly, and thus for b_2 small enough, the change from the field $V_2(x,y)$ can easily be made less than $1/4$ rad in direction (with the vectors normalized, of course). Furthermore, for $b_2(U_1,U_2)$ small enough, the longest are passing through a region of uniqueness in the remodified region is less that half of what is was before, or less than $1/6$. We can, of course, do this same modification for each strip U_n to give us a modification of all of U_1, and then do that for each strip U_m, by choosing $b_2(U_m,U_n)$ small enough. The new field we call $V_3(x,y)$.

We continue in this manner. For each triple (U_m,U_n,U_k) of strips, we define a number $b_3(U_m,U_n,U_k)$ so that when we modify first by applying a b_3 modification to U_k, followed by a $b_2(U_m,U_n)$ modification to U_n, followed by a $b_1(U_m)$ modification of U_m, the result differs from $V_3(x,y)$ in only a small region, differing there by no more than $1/8$ rad and cutting the length of the longest arc passing through a region of uniqueness to less than half of what it was before, i.e. to less than $1/12$.

We continue for infinitely many steps, and the field we obtain in the limit we denote as $V(x,y)$. The convergence is uniform, and reference to a sketch will show that every limiting vector lies in the first quadrant. It is clear that for each orbit, the longest arc passing through a region of uniqueness is arbitrarily short, i.e. the region of uniqueness is empty. The choice points are dense, and in fact, every point is a choice point. Through any point, and in every neighborhood of that point, there exist $\aleph$ flows whose velocities give the field $V(x,y)$ whose orbits through the given point disagree in pairs within the given neighborhood.

4.2 THEOREM: Let $V(x,y)$ be the field defined in $\mathbb{R}^2$ by Example 4.1. Then we can find two flows φ and Ψ in $\mathbb{R}^2$ such that

i) $\forall (x,y) \in \mathbb{R}^2$, $\mathcal{O}_{\varphi}(x,y) \cap \mathcal{O}_{\Psi}(x,y) = \{(x,y)\}$.

ii) $\forall (x,y) \in \mathbb{R}^2$, $\dot{\varphi}(x,y) = \dot{\Psi}(x,y) = V(x,y)$.

PROOF: We let m_1 be any measure on K whose total measure is no more than $1/2$, and such that for any open interval I which intersects K, $m_1(I \cap K) > 0$. Using the notation in Example 3.1, we see that the curves $y = f(x+C)$ are the orbits of a flow whose velocities are the field $V_1(x,y)$, and the curves $y = f_{m_1}(x+C)$ are also such orbits. Given any two of these curves, they intersect in an arc which lies in the region of uniqueness of the field $V_1(x,y)$.

We now define a measure m_2 somewhat similarly with m_1. For each component interval U of $\mathbb{R} \setminus f(K)$, we construct a measure $m_2(U)$ in such a way that the total of all these measures for all such U adds up to no more than $1/4$. Then, for each such U, we look at the curves $y = f(x+C)$ and $y = f_{m_2(U)}(x+C)$ in the strip where $0 \leq x \leq 1$. Compressing this strip in the horizontal direction by the width of C, and in the vertical direction by $b_1(U)$, we see that these curves are solution curves of the field generated in the same way in the construction of Example 4.1. Now adding to all these curves the required function $(y = f^{-1}(x), \forall x \in f(U))$ as before, we then retranspose, and obtain new solutions for the field $V_2(x,y)$. Using these curves to cross the strips U, we can put them together to give solutions to the field $V_2(x,y)$, in two ways, depending on whether we use the $m_2(U)$ or not. In addition, we can use m_1 or not, as we choose. Thus, we see the possibility for four different sets of curves, depending whether we use m_1 alone, or m_2, or neither, or both, and all of these have for their tangents the field $V_2(x,y)$. Actually, we will not be using the measure m_2 for certain technical reasons; its construction is included for pedagogical reasons. The reason for not using m_2 is that while m_1 has the effect of forcing solution curves to the right as they rise, so that modified curves increase more slowly than unmodified ones, m_2 has the effect of pushing them upward, so that it does not re-enforce the effect of m_1, but might possibly undo it.

We continue, however. For each pair U_1, U_2 of component intervals of $\mathbb{R} \setminus f(K)$, we choose a measure $m_3(U_1,U_2)$ so that the total of all these measures for all pairs does not exceed $1/8$, and also so that for each pair, and for each open interval I which intersects K, $m_3(U_1,U_2)(I \cap K) > 0$. Using the total measure m_3 in the same way as indicated before to obtain modified solutions for $V_3(x,y)$, we see that if we compare the solutions which involve both m_1 and m_3, in the indicated manner, to a solution which involves neither m_1 nor m_3, the orbits of the one solution intersect the orbits of the other in arcs each of which is contained in the region of uniqueness of $V_3(x,y)$, or else in a single point.

We continue in this manner, constructing for each integer n the configuration m_n of measures, where the total of measures at the nth stage is no more than 2^{-n}, and where the application of the measures in m_n force the solutions of $V_n(x,y)$ to the right if n is odd, and up if n is even. Applying only the odd-numbered measures, we find that the modified solutions are being forced more and more to the right by applications of the measures, though corresponding points are never further than a distance 1 apart. Against this, we consider solutions which are totally without the application of any measures at all. We see that the orbits of the two flows must intersect in arcs which lie in all the regions of uniqueness of $V_n(x,y)$ for all odd n. Thus, two such orbits intersect in a single point. QED

4.3 COROLLARY: Let $V(x,y)$ be the field defined in $\mathbb{R}^2$ by Example 4.1.

Then we can find a one-parameter family φ_α of flows such that

i) for each α, $\dot{\varphi}_\alpha(x,y) = V(x,y)$, $\forall\ (x,y) \in \mathbb{R}^2$,

and ii) $\forall\ (x,y) \in \mathbb{R}^2$, the orbits $\mathcal{O}_{\varphi_\alpha}(x,y)$ are pairwise disjoint except for the point (x,y).

PROOF: For each non-negative number α, make the construction shown in Theorem 4.2, using the measures αm_n instead of m_n. Now the same considerations show that for any two different values of α, the orbits are non-intersecting except at the initial point (x,y). QED

4.4 REMARK: The examples given in Corollary 4.3 represent an advance over the example of Hartman cited earlier. That example, though only for solution curves (not flows), did not have the strong separation of solutions obtained here.

4.5 PARADOX The orbits of φ and ψ in Theorem 4.2 have the following properties :

i) Each orbit is a differentiable curve.

ii) Each point of each φ-orbit is a crossing-path of a ψ-orbit and conversely.

iii) At the crossing-point, the two orbits are tangent.

It is interesting to note that this paradox grossly offends the intuition. This author finds it impossible to visualize the situation described, yet it is clearly true. Furthermore, dilligent attention to detail, specifically careful attention to the choice of the constants b_i, will even assure that the curves are C^∞. We can even refine this paradox:

4.6 PARADOX: There exists a homeomorphism h of $\mathbb{R}^2$ onto itself which carries the lines $y = C$ and $x = C$ into two families of differentiable curves with the property that the crossing points are points of tangency, and these, of course are all points.

To show this, let us look again at the orbits under φ and ψ in Corollary 4.3. Let us take the φ-orbit through the origin and consider the arc of length 2 with the origin at the center. We consider the family $\mathfrak{O}$ of ψ-orbits which intersect this arc, and consider an arc of length 2 on each of these orbits, centered at the intersection point.

Let us now attend to the ψ-orbit through the origin. For $\varepsilon > 0$ chosen small enough, every point on the indicated ψ-orbit has the property that the φ-orbit through it intersects all of the arcs just constructed. Let us assume that ε is chosen to satisfy this condition, and assume WoLOG that $\varepsilon = 1$. Now consider all the intersections of orbits passing through the indicated arcs on the φ-orbit and the ψ-orbit through 0. Let the mapping h_0 map them in the obvious way onto the set of all points $(x,y) \in \mathbb{R}^2$ with $|x| < 1$, $|y| < 1$, such that each orbit arc is mapped into a line of the form $x = C$ or $y = C$. We are almost done. The set of orbit intersections is bounded, and its interior is a bounded open disc. Let h_1 be a conformal mapping of that set onto the disc of radius $\frac{\pi}{2}$ about 0 in the complex plane. Now defining two maps T and E in R^2 by $T(x,y) = (\frac{2}{\pi}\tan^{-1}(x), \frac{2}{\pi}\tan^{-1}(y))$ and $E(r\,e^{i\theta}) = (\tan r)e^{i\theta}$, we easily see that the mapping $h = Eh_1h_oT$ has the required property.

5. GENERAL CONSIDERATIONS It would be instructive at this point to consider the whole of Example 4.1 as a modification of uniform horizontal flow. To do so, we introduce a contraction in the y-direction by a constant b_o. If b_o is chosen small enough, then we see that the modification is as small as we might wish. Thus, if a vector field V_o has a portion where the vectors are all parallel, or better yet all equal and non-zero, then we can modify that vector field as little as we please to produce a new field for which the given region is one where the new field has no local uniqueness. Do we need so stringent a condition? Clearly not. It suffices that the vectors in the indicated region be nearly equal. Whatever "nearly" means in the previous sentence, it is satisfied for each point of $\mathbb{R}^2$ by the vectors in some neighborhood of that point, if the vector field V_o is continuous. This sketch indicates the proof of the following theorem:

5.1 THEOREM: Let V_o be any continuous vector field in $\mathbb{R}^2$, and let U be any region disjoint from the set of points where V_o vanishes.

Then for each $\varepsilon > 0$, there exists a continuous vector field V_ε such that

i) V_ε has no local uniqueness in U for its flow solutions, and

ii) $||V_o(x,y) - V_\varepsilon(x,y)|| < \varepsilon, \quad \forall\ (x,y) \in \mathbb{R}^2$.

Written somewhat differently, the same theorem asserts

5.2 THEOREM: Let C^x be that subset of C^o (the set of continuous bounded vector fields in the plane) for which the flow solutions have no local uniqueness at any points where the field does not vanish.

Then C^x is dense in C^o in the uniform norm of C^o.

Compare this theorem with the following, taken from Choquet [1] vol. 1 pp. 121 ff:

5.3 THEOREM: The subset of C^0 which fails of uniqueness (even global uniqueness) at even a single point of $\mathbb{R}^2$ is a set of first category in C^0 in the uniform norm of C^0.

If the work above seems to leave the world of differential equations too chaotic a place, let me offer my own version of the spirit of Hope left at the bottom of Pandora's Box. My work in the area leaves me convinced (without proof) of the truth of the following assertion:

5.4 CONJECTURE: Let φ and ψ be two continuous velocity flows in $\mathbb{R}^2$ satisfying the following conditions:

i) $\varphi(x,y) = \psi(x,y), \ \forall (x,y) \in \mathbb{R}^2$.

ii) Each φ-orbit is analytic (either a point or a conformal image of the unit circle or the conformal image of the real axis).

iii) Each ψ-orbit is analytic.

Then $\varphi = \psi$.

Note that this must combine the properties of flows with those of analytic curves. If we relax the condition of analyticity to C^∞, we have the counterexamples shown here. If we reduce the condition of flow solutions, then Example 1.3 is a counterexample. The question is harder than it looks at first.

6 QUALITATIVE CONSIDERATIONS One of the interesting things about all the flows considered so far is that they have all been parallelizable. That is, they have all been conjugate, by a homeomorphism of the plane to the unique flow which is the solution of the differential equation $V(x,y) = (1,0)$. It is conceivable that all such bad examples are more or less the same, when it comes to

homeomorphic equivalence, or at least to the equivalence known as <u>orbit conjugacy</u>:

6.1 DEFINITION: The flow φ defined in the space X is said to be <u>orbit-conjugate</u> to the flow ψ defined in the space Y if there is a homeomorphism h of X into Y such that $h(\mathcal{O}_\varphi(x)) = \mathcal{O}_\psi(h(x))$, $\forall x \in X$. This is one of the weakest equivalence relations between flows, and often satisfied.

6.2 EXAMPLE: Let φ be the flow constructed, with that name, in Example 4.1, and for each measure $m \in M(K)$, we will consider the flow φ_m. Let us now consider the mapping $\exp_i$ defined in the complex plane by $\exp_i(z) = e^{iz}$ and for each measure $m \in M(K)$, define a flow ψ_m by setting $\psi_m(t, \exp_i(z)) = \exp_i(\varphi_m(t,z))$, $\forall t \in \mathbb{R}$, $z \in \mathbb{C}$, and $\psi_m(t,0) = 0$, $\forall t \in \mathbb{R}$. There is no ambiguity in this definition, since whenever we have $\exp_i(z_1) = \exp_i(z_2)$, z_1 must differ from z_2 by an integral multiple of 2π, so that, for any t, $\varphi_m(t, z_1)$ differs by the same quantity from $\varphi_m(t, z_2)$, and so $\exp_i(\varphi_m(t,z_1)) = \exp_i(\varphi_m(t,z_2))$. In examining the orbits of ψ_m, we observe that 0 is the only fixed point, that the circle of radius r about 0 is an orbit exactly if $\ln(r)$ has the property that for every interval I containing $\ln(r)$, $m(I) = \infty$, and all the other orbits spiral outward between two of these, with the exception that the origin of the spiral might be 0, and its terminus might be ∞.

The set of circle orbits, taken with 0 and ∞, is nowhere dense and closed. No more can be said, up to homeomorphism, for the description given here is exhaustive. For any closed, nowhere dense set F in $[0,\infty]$ which includes 0 and ∞, we can find a measure $m \in M(K)$ such that the flow ψ_m described above is orbit-conjugate with any flow whose periodic orbits are the circles $|z| \in F$, where these are traversed in the same direction, and all other orbits spiral outward between two of these. We see at once that all the flows Ψ_m are continuous velocity flows, and that they have the same velocities at every point. They are not, however, non-singular; the sole singularity is at 0. However, there are plenty of them, in fact plenty of conjugate classes. There is, in fact, one such class for every homeomorphic class of closed, nowhere dense sets in the closed interval, i.e. there are $\aleph$ such classes. For the same sort of thing with no singularities, see the next example.

6.3 EXAMPLE: We will now build another example; in this case, it will be a family of non-singular continuous velocity flows with the same velocities with no two being orbit-conjugate. To do this, we will first construct in the strip $\{(x,y) \mid -2 \leq y \leq -1\}$ a vector field like that constructed in the strip $\{(x,y) \mid 0 \leq y \leq f(1)\}$ in Example 3.1. In the strip $\{(x,y) \mid 1 \leq y \leq 2\}$, we do the same thing, except that we reverse the ends of the strip, so that the vectors are in the second quadrant instead of the first. For (x,y) between -1 and 1, we take the tangents to the semicircles $x-C = \sqrt{1 - y^2}$, i.e. the vector at the point (x,y) is $(-y, \sqrt{1 - y^2})$, where the sign of the radical is always positive.

If we apply to the two strips measures which are infinite in every neighborhood of -2 or $+2$ but finite in the neighborhood of every other point, we will get solution curves which have $y = -2$ and $y = +2$ as asymptotes and which increase, changing direction as they go. Such a strip cannot be mistaken for a parallel flow strip, for when the boundary orbits are included, the orbit space is not Hausdorff. If we take measures which are infinite only in the neighborhoods of -2, $+2$, and one other negative point, call it y^-, then the solution curves will give us a strip just like the one mentioned above in the strip between $y = y^-$ and $y = 2$, and in the strip between $y = -2$ and $y = y^-$, a parallel flow. If the points of infinitude are -2, $+2$, and a positive number y^+, then we will get a similar thing, with the parallel strip above the order-reversing strip, and if we have infinitude at -2, $+2$, y^-, and y^+, then we will have the order-reversing strip between two parallel strips.

We use these observations in connection with a vector field constructed as follows: In the strip $\{(x,y) \mid 0 \leq y \leq 4\}$, we use the vector field which we have just constructed, moved up by 2. In the strip $\{(x,y) \mid 4 \leq y \leq 8\}$, we use the same strip, but inverted, so that the edges match. Then in $\{(x,y) \mid 8 \leq y \leq 12\}$, we invert again, so that the strip regains its original orientation, and continue, withe the same vector field but inverted in the odd-numbered strips. For the lower half-plane, we use the constant vector field $(1, 0)$.

Each solution of this differential equation will exhibit a parallel flow in the lower half-plane, and infinitely many orientation-reversing strips in the upper half-plane. Between any two of these strips, we will find either a parallel region or a single orbit. These situations are topologically distinguishable. For each solution, we can construct a sequence of 0s and 1s in the following way: the first symbol is 0 if the first two strips are divided by a line, 1 if they are divided by a parallel strip, the second symbol is 0 or 1 according as the second and third strips are divided by a single orbit or a parallel region, <u>etc</u>. For each sequence of 0s and 1s, we can find a solution which will generate that sequence. Two solutions which generate the same sequence are orbit-conjugate; otherwise they are not. There are $\aleph^{s}$ such sequences. Therefore there are $\aleph^{s}$ different orbit-conjugacy classes of solutions. Note that the field constructed was a continuous field of unit vectors.

BIBLIOGRAPHY

Gustave Choquet (with the assistance of J. Marsden, T. Lance, and S. Gelbart)

[1] Lectures on Analysis, W. A. Benjamin, New York, 1969

Earl A. Coddington and Normal Levinson

[1] Theory of Ordinary Differential Equations, McGraw-Hill, New York, 1955.

Philip Hartman

[1] Ordinary Differential Equations, Wiley, New York, 1964.

SYMBOLIC DYNAMICS FOR HYPERBOLIC SYSTEMS

Rufus Bowen
University of California, Berkeley

Let $\mathfrak{M}$ be a finite set (called symbols) with the discrete topology and $\Sigma_{\mathfrak{M}} = \prod_{Z} \mathfrak{M}$ be the set of all doubly infinite sequences of symbols with the product topology. We shall be interested in the homeomorphism $\sigma : \Sigma_{\mathfrak{M}} \to \Sigma_{\mathfrak{M}}$ which shifts a sequence one space to the left. In symbolic dynamics one studies a differentiable system (diffeomorphism or flow) by relating it to the shift homeomorphism σ. This idea started with Hadamard [11] in 1898 when he set up a correspondence between geodesics on a surface of negative curvature with certain symbolic sequences. Then Marston Morse used this correspondence to construct a nonperiodic almost periodic geodesic [15], [16], [17]. Over the years Professor Hedlund has been active in symbolic dynamics (see for example [12] and [13]).

Today we shall outline the symbolic dynamics of the Axiom A diffeomorphisms and flows of Smale [22]. We were motivated by several earlier examples: Smale's horseshoe [21], [22], Adler and Weiss [1] on automorphisms of the 2-torus, and Sinai's Markov partitions for Anosov diffeomorphisms [20].

We first consider a diffeomorphism $f: M \to M$ on a compact Riemannian manifold. A closed f-invariant subset Λ of M is called <u>hyperbolic</u> if $T_\Lambda M$, the tangent bundle of M restricted to Λ, has continuous subbundles E^u and E^s which are invariant under Df such that $T_\Lambda M = E^u + E^s$ and there are constants $c > 0$ and $1 > \lambda > 0$ so that

$$\|Df^m(u)\| \leq c\lambda^m\|u\| \quad \text{for} \quad u \in E^s, \quad m \geq 0$$

and

$$\|Df^{-m}(v)\| \leq c\lambda^m\|u\| \quad \text{for} \quad v \in E^u, \quad m \geq 0.$$

The nonwandering set $\Omega(f)$ is defined by

$$\Omega(f) = \{x \in M: \text{ for every neighborhood } U \text{ of } x, \quad f^n(U) \wedge U \neq \phi \quad \text{for some } n > 0\}.$$

We say that f satisfies <u>Axiom A</u> if

(a) $\Omega(f)$ is hyperbolic

and (b) the periodic points of A are dense in $\Omega(f)$.

Smale's spectral decomposition [22] states that under these conditions there is a unique way of writing $\Omega(f) = \Omega_1 \cup \ldots \cup \Omega_n$ where the Ω_i are disjoint, closed, invariant subsets with each $f|\Omega_i$ topologically transitive. We call $f: \Omega_i \to \Omega_i$ (or Ω_i) a <u>basic hyperbolic set</u>. The set Ω_i may be very complicated; we use symbolic dynamics to describe $f: \Omega_i \to \Omega_i$.

We now return to symbols. Suppose $A: \mathfrak{M} \times \mathfrak{M} \to \{0,1\}$ and define

$$\Sigma_A = \{(m_i)_{i=-\infty}^{+\infty} \in \Sigma_{\mathfrak{M}}: A(m_i, m_{i+1}) = 1 \quad \text{for all } i \in Z\}.$$

Then Σ_A is a closed σ-invariant subset of $\Sigma_{\mathfrak{M}}$. We call $\sigma: \Sigma_A \to \Sigma_A$ (or Σ_A) a <u>subshift of finite type</u> if $\sigma|\Sigma_A$ is topologically transitive. Now it turns out that such subshifts of symbol sequences occur as basic hyperbolic sets [22], [23]. In fact the main idea of our symbolic dynamics is that they are the <u>model</u> basic sets.

Theorem 1 [2], [3]. Let $f:\Omega_i \to \Omega_i$ be a basic hyperbolic set. Then there is a subshift of finite type $\sigma:\Sigma_A \to \Sigma_A$ and a continuous surjection $\pi:\Sigma_A \to \Omega_i$ such that

(a) $\pi\sigma = f\pi$

(b) π is bounded finite to-one and is one-to-one over a Baire set of Ω_i

and (c) $x \in \Sigma_A$ is periodic, transitive, recurrent, or almost periodic (w.r.t. σ) if and only if $\pi(x)$ is (w.r.t. f).

One can use π to study the almost periodicity for $f:\Omega_i \to \Omega_i$, much as Morse did.

Theorem 2 [24]. Suppose Ω_i is an infinite set. Then the nonperiodic almost periodic points of f are dense in Ω_i.

Theorem 3 [3]. The minimal sets for $f:\Omega_i \to \Omega_i$ are precisely the sets $\pi(Y)$ where Y is minimal for $\sigma:\Sigma_A \to \Sigma_A$. All the minimal subsets for f are zero-dimensional.

Remark. Because $f|\Omega_i$ is expansive the last result implies that a minimal subset for $f|\Omega_i$ is topologically equivalent to some subshift.

Another thing one can study is periodic orbits. Let $N_n(f)$ be the number of fixed points of f^n. To study the numbers $N_n(f)$ one introduces the zeta function

$$\zeta_f(t) = \exp \sum_{m=1}^{\infty} (1/m) N_m(f) t^m .$$

Smale [22] conjectured that $\zeta_{f|\Omega_i}$ would be a rational function. J. Guckenheimer [10] proved this with an additional weak restriction on f. Bowen and Lanford [8] proved $\zeta_{\sigma|\Sigma_A}$ was rational for Σ_A a subshift of finite type. This led me to try to use symbolic dynamics to get at ζ_f. A. Manning proved the theorem.

Theorem 4 (Manning [14]). $\zeta_{f|\Omega_i}$ is rational.

Another look at periodic points leads one into ergodic theory. It turns out that the periodic points of $f|\Omega_i$ are equidistributed with respect to a Borel probability measure μ_f on Ω_i [6]. This means that if $N_n(U)$ is the number of fixed points of $f^n|\Omega_i$ in U and $\mu_f(\partial U) = 0$, then

$$\frac{N_n(U)}{N_n(f)} \to \mu_f(U) \quad \text{as} \quad n \to \infty.$$

The measure μ_f is f-invariant and in fact ergodic [6]. Ergodic properties of μ_f are to be seen as statements about asymptotic behavior of periodic points of f. Now $\sigma : \Sigma_A \to \Sigma_A$ is a basic set itself and so there is a μ_σ.

Theorem 5 [2]. The automorphisms (σ, μ_σ) and (f, μ_f) of measure spaces are conjugate by π.

Now μ_σ turns out to be a measure studied by Parry [19] earlier and μ_f is Haar measure when f is a hyperbolic automorphism of a torus [6]. On the 2-torus then theorem 5 is due to Adler and Weiss [1]. Now Parry [19] showed (σ, μ_σ) is a Markov chain, so (f, μ_σ) is as well. Using the recent work of Friedman and Ornstein [18] we get

Theorem 6 [2]. If (f, μ_f) is mixing, then it is Bernoulli.

Remark. (f, μ_σ) being mixing is equivalent to a certain topological condition on $f|\Omega_i$ [2].

A recent invariant of ergodic theory is entropy. For any f-invariant probability measure ν on Ω_i there is an entropy $h_\nu(f|\Omega_i)$. There is also a topological entropy $h(f|\Omega_i)$. A general theorem of Goodwyn [9] implies that $h_\nu(f|\Omega_i) \leq h(f|\Omega_i)$. Parry [19] proved that μ_σ was the only measure ν on Σ_A with $h_\nu(\sigma|\Sigma_A) = h(\sigma|\Sigma_A)$. Using the map π one gets

<u>Theorem 7 [2]</u>. μ_f is the only f-invariant probability measure ν on Ω_i with $h_\nu(f|\Omega_i) = h(f|\Omega_i)$.

The author feels that this topological characterization is particularly pleasing esthetically. It vaguely resembles an idea from statistical mechanics. D. Ruelle [25] has made this connection less vague.

We now turn to flows. The definition [22] of hyperbolicity and Axiom A is the same except that the splitting of $T_\Lambda M$ has an additional factor E ($T_\Lambda M = E + E^s + E^u$) where E is generated by the derivative of the flow. The model gets a bit more complicated. Let Σ_A be a subshift of finite type and $\alpha > 1$. Define the metric d_α on Σ_A by

$$d_\alpha((x_i),(y_i)) = \alpha^{-N}$$

where N is the largest integer such that $x_i = y_i$ for all $|i| < N$. Now assume that g is a positive real valued function on Σ_A which is Lipschitz with respect to d_α for some $\alpha > 1$. Define $Y \subset \Sigma_A \times R$ by

$$Y = \{(x,t) : t \in [0,f(x)]\} \quad .$$

Identifying $(x,f(x)) \sim (\sigma(x),0)$ one gets a metric space

$\Lambda(\sigma,f) = Y/\sim$. One can get a flow $\Psi = \{\psi_t\}$, the <u>suspension</u> of σ under f, by

$$\psi_s(x,t) = (x,t+s)$$

for small t and s and then using identifications to get a flow on $\Lambda(\sigma,f)$. We call Ψ on $\Lambda(\sigma,f)$ (with the above restriction on f) a <u>hyperbolic symbolic flow</u> [5].

Using this model one can prove the analogues of theorems 1, 2, and 3; now all minimal sets are one-dimensional [24]. If $\Phi = \{\phi_t : \Omega_i \to \Omega_i\}$ is a basic set, the periodic orbits are again equidistributed with respect to a measure μ_Φ and theorem 5 is again true [4]. Theorem 7 is also true [7], but the "model" case is no simpler than the general case; symbolic dynamics doesn't help us here. The trouble here is that the model is too complicated. The analogues of theorems 4 and 7 reduce to the case of our model [24] but are not yet known for the model.

<u>Problem 1</u>. If (ψ_t, μ_Ψ) is weak mixing (on $\Lambda(\sigma,f)$), is every ψ_t is Bernoulli.

<u>Remark</u>. Recent work of Ornstein suggests this problem.

For the zeta function one defines [22]

$$Z_\Phi(s) = \prod_{\gamma\in\Gamma} \prod_{k=0}^{\infty} (1 - [\exp \ell(\gamma)]^{-s-k})$$

where Γ is the set of closed orbits of $\phi_t : \Omega_i \to \Omega_i$ and $\ell(\gamma)$ is the minimal period of γ. Smale [22] asks whether $Z_\Phi(s)$ has a meromorphic continuation to the whole complex plane. Symbolic dynamics [24] reduces this to the problem for the model.

<u>Problem 2</u>. Does $Z_\Psi(s)$ have a meromorphic continuation to the plane?

We mention finally one problem that might be reasonable for diffeomorphisms.

Problem 3. Use symbolic dynamics to classify all basic sets $f|\Omega_i$.

References

1. R. L. Adler and B. Weiss. Entropy, a complete metric invariant for automorphisms of the torus, Proc. Natl. Acad. Sci. 57

2. R. Bowen. Markov partitions for Axiom A diffeomorphisms, Amer. J. Math. 92 (1970), 725-747.

3. _______ . Markov partitions and minimal sets for Axiom A diffeomorphisms, Amer. J. Math. 94 (1970), 907-918.

4. _______ . Periodic orbits for hyperbolic flows, Amer. J. Math.

5. _______ . One-dimensional hyperbolic sets for flows, Jour. Diff. Eqns.

6. _______ . Periodic points and measures for Axiom A diffeomorphisms, Trans. Amer. Math. Soc. 154 (1971), 377-397.

7. _______ . Maximizing entropy for a hyperbolic flow, Math. Systems Theory.

8. R. Bowen and O. E. Lanford. Zeta functions of restrictions of the shift transformation, Proc. Symp. Pure Math., Vol. 14 (1970), 43-49.

9. L. W. Goodwyn. Topological entropy bounds measure theoretic entropy, Proc. Amer. Math. Soc. 23 (1969), 679-688.

10. J. Guckenheimer. Axiom A and nocycles imply $\zeta_f(t)$ rational, Bull. Amer. Math. Soc. 76 (1970), 592-594.

11. J. Hadamard. Les surfaces a courbures opposees e leur liques geodesiques, Jr. Math. Pures Appl. 4 (1898), 27-73.

12. G. A. Hedlund. Endomorphisms and automorphisms of the shift dynamical system, Math. Systems Theory, 3 (1969), 320-375.

13. G. A. Hedlund and M. Morse. Symbolic dynamics, Amer. J. Math. 69 (1938), 815-866.

14. A. Manning. Axiom A diffeomorphisms have rational zeta functions, functions, Bull. London Math. Soc. 3 (1971), 215-220.

15. M. Morse. Representation of geodesics, Amer. J. Math. 43 (1921), 33-51.

16. _________. Recurrent geodesics on a surface of negative curvature, Trans. Amer. Math. Soc. 22 (1921), 84-110.

17. _________. Symbolic dynamics, lecture notes, Institute for Advanced Study, Princeton.

18. N. Friedman and D. Ornstein, On isomorphism of weak Bernoulli transformations, Advances in Math. 5(1970), 365-394.

19. W. Parry. Intrinsic Markov chains, Trans. Amer. Math. Soc. 112 (1964), 55-66.

20. Ya. G. Sinai. Markov partitions and C-diffeomorphisms, Func. Anal. and its Appl. 2 (1968), no. 1, 64-89.

21. S. Smale. Diffeomorphisms with many periodic points, "Differential and Combinatorial Topology", Princeton 1965, 63-80.

22. ________. Differentiable Dynamical Systems, Bull. Amer. Math. Soc. 73 (1967), 747-817.

23. R. Bowen. Topological entropy and Axiom A, Proc. Symp. Pure Math., Vol. 14 (1970), 23-41.

24. ________. Symbolic dynamics for hyperbolic flows, Amer. J. Math.

25. D. Ruelle, Statistical mechanics on a compact set with Z^{ν} action satisfying expansiveness and specification.

Monothetic Automorphisms of a Compact Abelian Group

James R. Brown*

In his paper [2] the present author introduced a condition on a compact abelian group G and an ergodic automorphism τ of G which guaranteed that τ was a factor of a certain "universal" automorphism. This condition is that there exist a point $a \in G$ whose orbit under τ separates the points of the dual group $\hat{G}$. It was pointed out that this condition (in fact a much stronger one) was always satisfied when τ was an ergodic automorphism of a compact abelian metric group. On the other hand, it is now known [5] that the condition is satisfied when $G = K^n$, the n-dimensional torus, and τ is any (not necessarily ergodic) automorphism. The purpose of the present paper is to give some rather general conditions, which include the above-mentioned examples, under which this condition is satisfied.

1. Definitions and Main Results. Throughout the paper G will be a compact abelian group and τ will be a continuous automorphism of G. We introduce now a condition equivalent to the one of the previous paragraph, which is, at

*Research on this paper was supported by NSF Contract GP-28944

the same time, a generalization of the well-known notion of a monothetic group introduced by van Dantzig. (See the footnote on p. 116 of [6]. Also see [7], [1] and [3].

Let us denote by $\mathcal{E}(\tau)$ the semigroup of endomorphisms ϕ of G of the form

$$\phi(x) = \sum_{j=1}^{\ell} n_j \tau^{k_j}(x),$$

where $n_1, \cdots, n_\ell, k_1, \cdots, k_\ell$ are integers and ℓ is a positive integer. If we denote by Gp Orb (a) the group generated in G by the orbit of a under τ, then clearly, Gp Orb (a) = $\mathcal{E}(\tau)a$.

<u>Definition</u> 1. The system (G,τ) is <u>monothetic</u> if $\exists a \in G$ with Gp Orb (a) = $\mathcal{E}(\tau)a$ dense in G.

<u>Remark</u>. If we denote the identity transformation on G by I, then (G,I) is monothetic iff G is a monothetic group in the usual sense. On the other hand, the following proposition relates this notion to the results of [2].

<u>Proposition</u> 1. <u>The system</u> (G, τ) is <u>monothetic iff</u> $\exists a \in G$ <u>such that</u> Orb (a) <u>separates the points of</u> $\hat{G}$.

<u>Proof</u>. Let G_o = Gp Orb (a). Then Orb (a) separates $\hat{G}$ iff the annihilator $G_o^{\perp}$ of the group G_o generated by

Orb (a) is trivial, i.e. iff $G_o^{\perp} = \{\hat{e}\}$, where $\hat{e}$ is the identity element of $\hat{G}$. This is true because the annihilator of a set A coincides with the annihilator of the group generated by A. On the other hand, the annihilator of the annihilator of G_o is the closure of G_o. Thus Orb (a) separates $\hat{G}$ iff $\overline{G}_o = G_o^{\perp\perp} = G$. ∎

It is well known [1, 3] that a group G is monothetic iff $\hat{G}$ is a subgroup of the circle group K_d with the discrete topology. Proposition 1 along with a slight modification of Theorem 2 and its Corollary of [2] yield the following analog of that result. We denote by $\bigotimes_{n=-\infty}^{\infty} K_d$ the product (complete direct sum) of countably many copies of K_d and by σ the shift transformation on $\bigotimes_{n=-\infty}^{\infty} K_d$.

Theorem 1. *The system* (G,τ) *is monothetic iff there is a shift-invariant subgroup* X *of* $\bigotimes_{n=-\infty}^{\infty} K_d$ *and a group isomorphism* ρ^* *of* G *onto* X *such that* $\rho^*\tau^* = \sigma\rho^*$.

The following theorem seems to be well-known. The first main result of the present paper (Theorem 3) is essentially the result of a closer analysis of its proof as given, for example, in [2].

<u>Theorem</u> 2. <u>If</u> G <u>is metrizable and</u> τ <u>is ergodic</u>, <u>then</u> $\exists a \in G$ <u>with</u> Orb (a) <u>dense in</u> G.

The analysis referred to above leads us to the following definition.

<u>Definition</u> 2. The system (G,τ) is <u>weakly topologically ergodic</u> (w. t. e.) if $A \subseteq G$, $\mathcal{E}(\tau)A \subseteq A$ implies A is dense or A is nowhere dense in G.

<u>Remark</u>. It is sufficient to verify the implication for closed sets A.

<u>Theorem</u> 3. <u>If</u> G <u>is metrizable and</u> (G,τ) <u>is weakly topologically ergodic, then</u> (G,τ) <u>is monothetic</u>.

<u>Proof</u>. Let $\{U_n\}$ be a countable base for the topology of G. For each n let

$$X_n = \{x \in G\colon\ U_n \cap \text{Gp Orb }(x) = \phi\}.$$

Suppose that $y \in \mathcal{E}(\tau)X_n$, say $y = \phi(x)$, where $\phi \in \mathcal{E}(\tau)$ and $x \in X_n$. Since Gp Orb(y) $\subseteq$ Gp Orb(x), it follows that $y \in X_n$. Thus $\mathcal{E}(\tau)\, X_n \subseteq X_n$. Since (G,τ) is w.t.e. and $X_n \cap U = \phi$, it follows that X_n is nowhere dense. Since the set of $x \in G$ for which Gp Orb(x) is not dense in G is $\bigcup_n X_n$, which is a first category set, it follows that there exist points x for which Gp Orb(x) is dense, that is (G,τ) is monothetic. ▮

Note that $\mathcal{E}(\tau)$ always contains I. Thus the following useful proposition is evident.

Proposition 2. *If (G,I) is w.t.e., then (G,τ) is w.t.e. for any τ. If (G,I) is monothetic (that is G is a monothetic group), then (G,τ) is monothetic.*

Remark. Note that (K^n,I) is w.t.e. For suppose that $A \subseteq K^n$, $\mathcal{E}(I)A = \bigcup_{k=-\infty}^{\infty} kA \subseteq A$, A is closed and contains an open "rectangle" B. Then $G \subseteq kB$ for sufficiently large k, and hence $A = G$. Thus (K^n,τ) is monothetic for any τ. A similar argument applies to the infinite-dimensional torus K^ω.

Proposition 3. *If τ is an ergodic automorphism of G, then (G,τ) is w.t.e.*

Proof. Suppose $A \subseteq G$, $\mathcal{E}(\tau)A \subseteq A$ and A is closed. Then $\tau A \subseteq A$. Since τ is ergodic, either $m(A) = 0$ or $m(\tilde{A}) = 0$, where m is Haar measure. Since $\tilde{A}$ is open, $m(\tilde{A}) = 0$ implies $\tilde{A} = \phi$, i.e. $A = G$. If $m(A) = 0$, then $A^o = \phi$, and since A is closed, it is nowhere dense. ▮

2. Examples. We offer some examples to illustrate the limitations of the results of the previous section as well as limitations on improving these results.

A. Let $G_1 = Z_2 \oplus Z_2$, where $Z_2 = \{0,1\}$ is a cyclic group of order 2. This group is not monothetic (i.e. not cyclic). In fact, if G is monothetic, then $\hat{G}$ cannot contain a subgroup isomorphic to $\hat{G}_1 = G_1$ [1,3].

Of course, (G_1,I) is not w.t.e. How about (G_1,τ) for other automorphisms τ of G_1? (There are no ergodic automorphisms, so Proposition 3 does not apply.) There are essentially only two automorphisms of G_1 other than the identity. The automorphism τ_1 interchanges $(1,0)$ and $(0,1)$ and leaves $(0,0)$ and $(1,1)$ fixed; τ_2 maps $(0,1)$ onto $(1,0)$, $(1,0)$ onto $(1,1)$, and $(1,1)$ onto $(0,1)$. It is readily verified that $\mathcal{E}(\tau_1) = \{0,I,\tau_1, I+\tau_1\}$ and $\mathcal{E}(\tau_2) = \{0,I,\tau_2,\tau_2^2\}$. Since $(0,0)$ is a fixed point for any endomorphism of G_1, neither (G_1,τ_1) nor (G_1,τ_2) is w.t.e. On the other hand, we see that $\mathcal{E}(\tau_1)\ (0,1) = \mathcal{E}(\tau_1)\ (1,0) = G_1$, while $\mathcal{E}(\tau_1)\ (1,1) = \{(0,0),\ (1,1)\}$, and that $\mathcal{E}(\tau_2)x = G_1$ for each $x \neq (0,0)$. Thus (G_1,τ_1) and (G_1,τ_2) are monothetic.

B. Let $G_2 = K \oplus Z_2$. If $a = (w,1)$, where w is irrational, then $Gp(a)$ is dense in G_2. Thus (G,I) and hence any (G,τ) are monothetic. On the other hand, (G,I) is not w.t.e. For if $A = K \oplus \{0\}$, then $\mathcal{S}(I)A \subseteq A$, but A is neither dense nor nowhere dense. In fact, no automorphism of the form $\tau = \phi \oplus \psi$ is w.t.e. Thus the converse if Theorem 3 fails badly.

C. The group $\hat{K}_d$ is, of course, monothetic. The group $G_3 = \hat{K}_d \oplus \hat{K}_d$ is not. There are, however, monothetic automorphisms of G_3. For example, let $\tau_3(x,y) = (y,x)$. If a_o is a topological generator of $\hat{K}_d$, then $a = (a_o,0)$ has dense group orbit. Indeed, the set of elements of the form

$$n_1 a + n_2 \tau_3(a) = (n_1 a_o,\ n_2 a_o)$$

is dense in G_3. Theorem 3 does not apply since G_3 is not metrizable.

D. Let $G_4 = \bigotimes_{n=-\infty}^{\infty} \hat{K}_d$ be the direct product of countably many copies of $\hat{K}_d$, and let τ_4 be the shift operator $(\tau_4 x)_n = x_{n-1}$. Then $\hat{G}_4 = \bigoplus_{n=-\infty}^{\infty} K_d$ (the direct sum or restricted product) is a shift-invariant subgroup of $\bigotimes_{n=-\infty}^{\infty} K_d$, and τ_4^* is the restriction of σ. According to

Theorem 1, (G_4, τ_4) is <u>monothetic</u>. Moreover, τ_4^* has no finite nonzero orbits. Hence τ_4 is <u>ergodic</u>. Again G_4 is <u>not metrizable</u>, so Theorem 2 and 3 do not apply.

3. <u>Algebraically monothetic groups</u>. Theorem 1 provides a characterization of those groups G for which there exists an automorphism τ of G and a point $a \in G$ such that Gp Orb(a) (under the action of τ) is dense in G, namely $\hat{G}$ must be isomorphic to a shift-invariant subgroup of $\bigotimes_{n=-\infty}^{\infty} K_d$.

<u>Definition</u> 3. The group G is <u>algebraically monothetic</u> if there exists an automorphism τ of G and $a \in G$ such that $\mathcal{E}(\tau)a = \text{Gp Orb}(a)$ is dense in G.

In this section we seek intrinsic conditions on G (or on $\hat{G}$) in order that it be algebraically monothetic. Note that every monothetic group is algebraically monothetic, but (Example A) not conversely. In fact, Example A typifies the difference between the two concepts as seen in the next two theorems.

In [1] and [3] is proven the following theorem.

<u>Theorem</u> 4. <u>The group</u> G <u>is monothetic iff the cardinality of</u> $\hat{G}$ <u>is no greater than</u> c (<u>the power of the continuum</u>)

and the torsion group $T(\hat{G})$ *of* $\hat{G}$ *is isomorphic to a subgroup of* K_d *(and hence of* $T(K_d)$*).*

It is further stated in [3] that the latter condition is equivalent to saying that every finitely generated subgroup of $T(\hat{G})$ is cyclic.

The key to the proof of Theorem 4 is showing that any monomorphism of $T(\hat{G})$ into K_d can be extended to a monomorphism of $\hat{G}$ into K_d. Precisely the same argument (transfinite induction on the elements of infinite order in $\hat{G}$) can be applied to show that any monomorphism of $T(\hat{G})$ onto a shift-invariant subgroup of $K_d^{\omega} = \bigotimes_{n=-\infty}^{\infty} K_d$ can be extended to a monomorphism of $\hat{G}$ onto a shift-invariant subgroup of K_d^{ω}. Thus we have the following theorem.

Theorem 5. *The group* G *is algebraically monothetic iff the cardinality of* $\hat{G}$ *is no greater than* c *and* $T(\hat{G})$ *is isomorphic to a shift-invariant subgroup of* K_d^{ω}.

From Theorem 4 we have the following corollary, which again includes the case $G = K^n$.

Corollary 4.1. *Every separable (and hence every metrizable) connected compact abelian group is monothetic.*

Proof. It is well known that G is connected iff $\hat{G}$ is torsion-free. Thus we need only show that the cardinality of $\hat{G}$ is no greater than c. Now if G is separable, then the continuous characters on G are determined by their values on a countable set of points in G. Hence there cannot be more than c of them. ∎

It is interesting to speculate whether or not every separable compact abelian group G is algebraically monothetic. We know of no counterexamples. We shall present some partial results in this direction. Note that while K_d contains only one cyclic subgroup of order p (so that Corollary 4.1 cannot be improved upon very much), K_d^{ω} contains uncountably many independent elements of order p.

It is known (see [4]) that the torsion group $T(G)$ is a countable direct sum of primary groups $T_p(G)$, where p ranges over the prime numbers. Thus the problem reduces to finding shift-invariant embeddings of $T_p(G)$ in $T_p(K_d^{\omega})$. Let us denote as usual by $Z(p^k)$ the cyclic group of order p^k consisting of elements of K_d of the form j/p^k $(j = 0,1,\cdots,p^k-1)$, and by $Z(p^\infty)$ the subgroup $Z(p^\infty) = \bigcup_{k=1}^{\infty} Z(p^k)$ of K_d. Then $Z(p^\infty) = T_p(K_d)$ and $T_p(K_d^{\omega}) = \bigotimes_{n=-\infty}^{\infty} Z(p^\infty)$.

Theorem 6. *If* $\hat{G}$ *is of the form* $\bigotimes_{n=-\infty}^{\infty} Z(p^k)$ *or* $\bigoplus_{n=-\infty}^{\infty} Z(p^k)$ *for some prime* p *and some* k, $1 \le k \le \infty$, *then* G *is algebraically monothetic*.

Proof. Obvious. ▮

Theorem 7. *If* $\hat{G}$ *is of the form* $\bigoplus_{j=1}^{s} Z(p^{k_j})$, *where* p *is prime and* $1 \le k_1 \le k_2 \le \cdots \le k_s \le \infty$, *then* G *is algebraically monothetic*.

Proof. We have $Z(p^{k_1}) \subseteq Z(p^{k_2}) \subseteq \cdots \subseteq Z(p^{k_s})$. Thus the mapping $\tau^*: \hat{G} \to \hat{G}$ is well-defined by

$$\tau^*(t_1,\cdots,t_\ell) = (t_1,t_1+t_2,t_2+t_3,\cdots,t_{\ell-1}+t_\ell).$$

It is readily verified that τ^* is an automorphism of $\hat{G}$. Now let $a \in G$ be defined by $\langle a,t\rangle = t_\ell$ for $t \in \hat{G}$. Here we use the notation $\langle a,t\rangle$ to denote the value of the character t at the point a.

Suppose now that t annihilates the orbit of a under τ. Then

$$\begin{aligned}\langle \tau^n a,t\rangle &= \langle a,\tau^{*n} t\rangle \\ &= \sum_{j=0}^{n} \binom{n}{j} t_{\ell-j}\end{aligned}$$

is zero for each $n = 0,1,\cdots,\ell-1$. It follows that $t = 0$. According to Proposition 1, (G,τ) is monothetic. ▌

Corollary 7.1. *If* $\hat{G}$ *is finitely generated, then* G *is algebraically monothetic.*

Proof. In this case the cardinality of $\hat{G}$ is less than c and $T_p(\hat{G})$ is of the form given in the theorem. Thus the embeddings for various values of p may be pasted together to give an embedding of $T(\hat{G})$, which may then be extended to an embedding of $\hat{G}$. ▌

Obviously, much more can be obtained from the last two theorems.

Theorem 8. *If* G *is separable and each* $T_p(\hat{G})$ *is of one of the forms mentioned in Theorems* 6 *and* 7, *then* G *is algebraically monothetic.*

It is known [4, p.10] that a divisible p-primary group is a direct sum of copies of $Z(p^\infty)$. Finite direct sums are covered by Theorem 7, countable direct sums by Theorem 6. The group $T(\bigotimes_{n=-\infty}^{\infty})Z(p^\infty)$ has cardinality c and is divisible and p-primary. This it must be a direct sum of c copies of $Z(p^\infty)$. It follows that such direct sums are also covered by Theorem 6. Thus we have the following

Corollary 8.1. *If* G *is separable and* $\hat{G}$ *is divisible, then* G *is algebraically monothetic.*

Since [4, p. 9] any group is a direct sum of a divisible group and a reduced group (i.e. having no divisible subgroups), the problem is now reduced to consideration of reduced p-primary groups. At least in the case where $\hat{G}$ is countable (but not finitely generated), we need to know whether all possible sets of Ulm invariants [4, p. 27] are represented among the reduced shift-invariant countable subgroups of $\bigotimes_{n=-\infty}^{\infty} Z(p^{\infty})$.

Note that it is not immediately obvious how to realize something as simple as a direct sum of countably many copies of $Z(p)$ and countably many copies of $Z(p^2)$ as a shift-invariant subgroup of K_d^{ω}. The same remark applies to $\bigoplus_{k=1}^{\infty} Z(p^k)$.

4. Generalizations. In this section we define a locally compact abelian T-module and give a generalization of Theorem 1. This generalization also includes the characterization of solenoidal groups by Anzai and Kakutani [1] and suggest an extension of their theorem which is analogous to Theorem 1. The point of view is that of [1] with the additional action of a group T.

Definition 4. Let T be an abstract group. The group H is a *locally compact abelian* T-*module* if (i) H is a locally compact abelian group, (ii) H is a (left) T-module, and (iii) the map $h \to th$ of H into H is continuous for each $t \in T$.

Definition 5. Let G and H be locally compact abelian T-modules. A mapping ϕ of H into G is a T-*homomorphism* if (i) ϕ is a continuous group homomorphism, and (ii) $\phi(th) = t\phi(h)$ for all $t \in T$, $h \in H$.

Theorem 9. *Let* G *and* H *be locally compact abelian* T-*modules with* G *compact*. *Let* $\bar{H}$ *denote the Bohr compactification of* H. *Then any* T-*homomorphism* ϕ *of* H *into* G *can be extended uniquely to a* T-*homomorphism of* $\bar{H}$ *into* G.

Proof. Let ϕ be such a homomorphism. It is well known that ϕ can be extended uniquely to a continuous group homomorphism $\bar{\phi}$ of $\bar{H}$ into G. Similarly, the action of T on H may be extended to $\bar{H}$. Indeed, $\bar{H}$ may be characterized by the fact that any continuous homomorphism of H into a compact group may be uniquely extended to $\bar{H}$, and the action of $t \in T$ on H may be thought of as a homomorphism of H into $\bar{H}$. Uniqueness then implies that $s(t\bar{h}) = (st)\bar{h}$ for all $s,t \in T$, $\bar{h} \in \bar{H}$. Similarly, since each of $\bar{\phi}(t\bar{h})$ and $t\bar{\phi}(\bar{h})$

is an extension to $\overline{H}$ of the homomorphism $\psi: H \to \overline{H}$ defined by $\psi(h) = \phi(th) = t\phi(h)$, they must be equal. Thus $\overline{\phi}$ is a T-homomorphism. ∎

Corollary 9.1. (Anzai-Kakutani). *Let* G *and* H *be as in the theorem. Then there exists* a T-*homomorphism* $\phi: H \to G$ *with* $\phi(H)$ *dense in* G *iff the dual* $\hat{G}$ *of* G *is algebraically isomorphic to a* (*not necessarily closed*) T-*invariant subgroup of* $\hat{H}$.

Proof. Consider the sequence of maps: $\phi: H \to G$, $\overline{\phi}: \overline{H} \to G$, $\overline{\phi}^*: \hat{G} \to \hat{\overline{H}} = \hat{H}_d$. The existence of ϕ implies the existence of $\overline{\phi}$ and $\overline{\phi}^*$. If $\phi(H)$ is dense in G, then $\overline{\phi}$ is epic. Hence $\overline{\phi}^*$ is monic. Conversely, existence of a monomorphism $\overline{\phi}^*$ obviously implies existence of $\overline{\phi}$ and hence ϕ with the desired property. ∎

Example 1. Let $T = \{e\}$ be the trivial group and $H = Z$. We obtain the Halmos-Samelson theorem about monothetic groups.

Example 2. Let T be trivial again, and let $H = R$ (the reals). This yields the Anzai-Kakutani characterization of solenoidal groups. Examples 1 and 2 as applications of Corollary 9.1 are contained in [1].

<u>Example</u> 3. Let $T = Z$ and $H = Z^{\infty} = \bigoplus_{n=-\infty}^{\infty} Z$.

The action of T on H is given by the powers of the shift: $t\{k_n\} = \{k_{n-t}\}$. The result is our Theorem 1 rewritten as a characterization of algebraically monothetic groups. Indeed, the most general form that $\phi: H \to G$ can take is $\phi(\{k_n\}) = \sum_n k_n \tau^n(a)$, where $a = \phi(\cdots,0,1,0,\cdots)$ and τ is an automorphism of G. Thus $\phi(H) = \text{Gp Orb}_\tau(a)$, and the result follows immediately.

<u>Example</u> 4. Let $T = R$ and $H = \bigoplus_{s \in R} Z_s$, $Z_s = Z$. The action of T on H is $t\{k_s\} = \{k_{s-t}\}$, and the resulting theorem is new. It is contained, as are Examples 1 and 3 in the following theorem.

<u>Theorem</u> 10. <u>In order that a compact abelian group</u> G <u>admit a group</u> $\sum = \{\sigma_t: t \in T\}$ <u>of continuous automorphisms and a point</u> $a \in G$ <u>such that the group generated by the</u> $\sum$<u>-orbit of</u> a <u>is dense in</u> G, <u>it is necessary and sufficient that</u> $\hat{G}$ <u>be algebraically isomorphic to a (not necessarily closed) shift-invariant subgroup of the torus</u> $K^T = \bigotimes_{s \in T} K$.

The proof is exactly as for Example 3, with $H = \bigoplus_{s \in T} Z$ and $\phi: H \to G$ given by

$$\phi(\{k_s\}) = \sum_s k_s \sigma_s(a).$$

This suggests, with Example 2, the following analog of the Anzai-Kakutani theorem on solenoidal groups. In this case we take $H = \bigoplus_{s \in T} R$ and look for the most general $\phi: H \to G$. By analogy with the above, we might expect this to take the form

$$\phi(\{h_s\}) = \sum_s h_s \sigma_s(a).$$

However, in general, multiplication by the real number h_s will not make sense in G. Moreover, this expression does not lead to a particularly easily stated description of $\phi(H)$. Instead, we note that each $h \in H$ may be written

$$h = \{h_s\} = \sum_s \bar{\sigma}_s(\bar{a}(h_s)),$$

where $\bar{a}(z)$ has all components zero except the zeroth one, which is z, and $\bar{\sigma}_s$ is the shift. Thus

$$\phi(h) = \sum_s \sigma_s a(h_s),$$

where $a: R \to G$ is a continuous homomorphism.

Theorem 11. *In order that a compact abelian group* G *admit a group* $\sum = \{\sigma_t : t \in T\}$ *of continuous automorphisms and a continuous automorphism* $a: R \to G$ *such that the* Σ*-invariant subgroup generated by* $a(R)$ *is dense in* G, *it is necessary and sufficient that* $\hat{G}$ *be algebraically isomorphic to a (not necessarily closed) shift-invariant subgroup of the dual of* $\bigoplus_{s \in T} R$. *If* T *is finite, this dual is* R^T. *If we drop the requirement that* a *be continuous, the condition is that* $\hat{G}$ *be isomorphic to a shift-invariant subgroup of* $\overline{R}^T = \bigotimes_{s \in T} \overline{R}$, *where* $\overline{R}$ *is the Bohr compactification of the reals*.

In the last case, we have taken for H the direct sum $\bigoplus_{s \in T} R_d$, where R_d is the reals with the discrete topology.

References

1. H. Anzai and S. Kakutani, Bohr compactifications of a locally compact group I, II. Proc. Jap. Acad. (Tokyo) 19(1943), 467-480, 533-539.

2. J. R. Brown, A Model for ergodic automorphism on groups. Math. Systems Theory (to appear).

3. P. Halmos and H. Samelson, On monothetic groups. Proc. Nat. Acad. Sci. U.S. 28(1942), 254-258.

4. I. Kaplansky, Infinite Abelian Groups, Univ. of Michigan Press, 1954.

5. J. D. Kerrick, Group automorphisms of the n-torus: a representation theorem and some applications. Ph.D. Dissertation, Department of Mathematics, Oregon State University, 1971.

6. D. Van Dantzig, Über topologisch homogene Kontinua, Fund. Math. 15(1930), 102-125.

7. D. Van Dantzig, Zur topologischen Algebra, Math. Annalen 107(1933), 591.

Department of Mathematics
Oregon State University
Corvallis, Oregon 97331

ON THE EMBEDDING PROBLEM AND THE HILBERT-SMITH CONJECTURE

Hsin Chu[1]

§1. INTRODUCTION.

In Topolgical Dynamics, the embedding problem on almost periodicity states as follows: "Given an effective, almost periodic, discrete flow (Z,M,π), where $Z = \{\phi^r | r = \text{integer}\}$, find a positive integer ℓ and a continuous flow (R,M,π') such that there is an isomorphism $f : Z' \longrightarrow R$, from Z' into R, where $Z' = \{(\phi^\ell)^r | r = \text{integer}\}$, $f(\phi^\ell) = 1$ and $\pi'(f(\phi^{\ell r}),x) = \pi(\phi^{\ell r},x)$." On the other hand, the Hilbert-Smith conjecture states that if a compact group acts effectively on a connected, n-manifold M, then the group is a Lie group. Some special cases of this conjecture were proved affirmatively. Among them, we know that the conjecture is true if $n = 1,2$ (See [6]). We do not know the case for $n = 3$. However, if (Z,M,π) has an orbit of dimension ≥ 1 and $n = 3$, the statement remains true (See [1]). In this note, we first show that if the space is a closed, connected 3-manifold, the solvability of the embedding problem is equivalent to the Hilbert-Smith conjecture. We then prove that if a compact group G acts effectively on a closed, connected 3-manifold M and if there exists a circle group S acting effectively on M such that every element in G commutes with every element in S, then G is a Lie group. We also show that if M is a closed, connected, 2-manifold and an effective almost periodic discrete flow (Z,M,π), where $Z = \{\phi^r | r = \text{integer}\}$ is given, then M must be either S^2, P^2, KL, or T^2, and there is a positive integer ℓ and an effective, almost periodic, continuous flow (R,M,π') such that (Z',M,π) can be embedded into (R,M,π') where $Z' = \{(\phi^\ell)^r | r = \text{integer}\}$. Our manifolds are always assumed to be separable.

---

[1]This research was partially supported by NSF Grant GP 29271.

§2. LEMMAS.

Let (T,M,π) be a transformation group, where M is compact Hausdorff. Define a continuous map f from T into M^M, by $f(t) = \{t(x) \mid x \in M\}$, for all $t \in T$, where M^M is the set of all maps from M into M with its usual cartesian product topology. Let $G = \overline{f(T)}$. If (T,M,π) is almost periodic then G is a compact topological group of homeomorphisms on M. Define a transformation group (G,M,π') as follows: $\pi' : G \times M \longrightarrow M$ by $\pi'(g,x) = g(x)$, for $g \in G$ and $x \in M$. The diagram:

$$\begin{array}{ccc} (T,M) & \xrightarrow{\ \pi\ } & M \\ f\big\downarrow \ \ i\big\downarrow & & i\big\downarrow \\ (G,M) & \xrightarrow{\ \pi'\ } & M \end{array}$$

is commutative, where i is the identity map. If (T,M,π) is effective, so is (G,M,π'). Thus we have:

<u>Lemma 1</u>. If (T,M,π) is an effective, almost periodic transformation group, where M is compact Hausdorff, then (G,M,π') is an effective, compact transformation group. Conversely, if (G,M,π') is an effective, compact transformation group then (T,M,π) is effective, almost periodic.

<u>Lemma 2</u>. Let M be a closed connected 3-manifold. Let G be a p-adic group. If there is a continuous flow (R,M,ρ) such that G and R have a common element which is not the identity, then G cannot act effectively on M.

<u>Proof</u>. Suppose the conclusion is false. Then G acts effectively on M. We denote this transformation group by (G,M,π). Let h be the given non-trivial common element of R and G. Define $H = \{h^n \mid n = \text{integer}\}$. Then H is a common subgroup of G and R. If (G,M,π) is effective, so is (H,M,π). We claim that R must act effectively on M. If R does not act effectively on M, then there

exists a closed subgroup Z of R such that R/Z acts on M effectively. Since R and G have a non-trivial common element, it is easy to see that $Z \neq R$ and Z must be an integer group in R. It follows that $R/Z = S^1$, a circle group. Let q be the natural projection from R onto $R/Z = S^1$. Since $H \cap Z = \{0\}$ and H cannot be a non-trivial finite group in the p-adic group G, we have $\overline{q(H)} = S^1$. We may embed S^1 and G into M^M by the natural maps f_1 and f_2 respectively, where $f_1(s) = \{s(x) | x \in M\}$ for $s \in S^1$ and $f_2(g) = \{g(x) | x \in M\}$ for $g \in G$. Then f_1 and f_2 are one-to-one continuous isomorphisms. It follows that f_1 is an isomorphism from S onto $f_1(S)$ and f_2 is an isomorphism from G onto $f_2(G)$. From $f_1 q(H) = f_2(H)$, we have $\overline{f_1 q(H)} = f_1(S) = \overline{f_2(H)} \subset f_2(G)$. This is impossible, because any non-trivial subgroup of a p-adic group G must be p-adic. Hence R acts effectively on M. By Lemma 1, H must act on M almost periodically. Since H is a syndetic subset of R, it is known from Toploogical Dynamics that R also acts on M almost periodically. By Lemma 1, there is a continuous isomorphism f which maps R into a compact group K in M^M such that $\overline{f(R)} = K$ and an effective transformation group (K, M, π'). Since R is connected so is K. Since $\dim M = 2$ the Hilbert-Smith conjecture is true and K must be a Lie group. However $f_2(h) = f(h)$, for each $h \in H$, in M^M, we have $f_2(H) \subset f_2(G) \cap K$. It follows that $\overline{f_2(H)} \subset f_2(G) \cap K$ and $\overline{f_2(H)}$ must be a non-trivial Lie group in a p-adic group $f_2(G)$. This is impossible. Hence G cannot act on M effectively.

§3. MAIN RESULTS.

Theorem 1. If the space is a closed, connected 3-manifold, the solvability of the embedding problem is equivalent to the Hilbert-Smith conjecture.

Proof. Suppose that the embedding problem can always be solved. Let (G, M, π) be an effective compact transformation group, where M

is the given space. If G is not a Lie group, it is known that G contains a p-adic group A_p for some prime p. The induced transformation group is also effective. Choose a non-trivial element $g \in A_p$ and let Z be the integer group generated by g. The transformation group (Z,M,π) must be effective and almost periodic. Then there exists an integer ℓ, a continuous flow (R,M,π') and an isomorphism h from Z into R such that the following diagram is commutative:

$$\begin{array}{ccc} (Z',M) & \xrightarrow{\ \pi\ } & M \\ {\scriptstyle h}\big\downarrow \; {\scriptstyle i}\big\downarrow & & \big\downarrow{\scriptstyle i} \\ (R,M) & \xrightarrow{\ \pi'\ } & M \end{array}$$

where $Z' = \{(g^{\ell})^r \mid r = \text{integer}\}$. Thus g^{ℓ} and $f(g^{\ell})$ are the same homeomorphism on M, and the group R and the group A_p have this non-trivial element in common. However, since by Lemma 2, (A_p,M,π) cannot be effective, this leads to a contradiction. Hence the group G must be a Lie group.

Conversely, suppose the Hilbert-Smith conjecture is true for this given space. Let (Z,M,π) be effective, almost periodic discrete flow. By Lemma 1, there exists a compact transformation group (G,M,π') and an isomorphism f from Z into G such that $\overline{f(Z)} = G$ with $f(\pi(\phi,x)) = \pi'(f(\phi),x)$ for every $x \in M$, where $\phi \in Z = \{\phi^n \mid n = \text{integer}\}$. The compact group G must be a Lie group. If G is discrete, then G must be finite and (Z,M,π) is not effective. Hence the connected component G_0 of the identity in G is non-trivial and G/G_0 is a finite group. It follows that there is an integer ℓ such that $f(\phi^{\ell})$ is non-trivial and $f(\phi^{\ell}) \in G_0$. There must exist an one-parameter subgroup R in G_0 such that $f(\phi^{\ell}) \in R$. Let $Z' = \{(\phi^{\ell})^n \mid n = \text{integer}\}$. We have $\pi(\eta,x) = \pi'(f(\eta),x)$ for $\eta \in Z'$ and $x \in M$. Hence the embedding problem is solvable.

Remark. In the proof of this theorem, we actually prove the following statement: "Let M be a closed, connected n-manifold for any n. If the Hilbert-Smith conjecture is true then the embedding problem can always be solved."

Theorem 2. If a compact group G acts effectively on a closed, connected 3-manifold M and furthermore, if there exists a circle group S acting effectively on M such that every element in G commutes every element in S, then G is a Lie group.

Proof. Suppose the statement is false. Let (G,M,π) be an effective transformation group, where G is compact but not Lie. Then it is known that G contains a p-adic group A_p, for some prime p. Again (A_p,M,π) is an effective transformation group. By the given effective, circle transformation group (S,M,π'), we define $(S\times A_p,M,\pi'')$ as follows: $\pi''((k,t),x) = \pi'(k,\pi(t,x))$ for $k\in S$, $t\in A_p$, $x\in M$. Since every element of S commutes with every element of A_p, it is not hard to see that $(S\times A_p,M,\pi'')$ is a transformation group. If it is not effective then there exist $k_0\in S$ and $t_0\in A_p$ such that $\pi''((k_0,t_0),x) = x$ for all $x\in M$, or $\pi'(k_0,\pi(t_0,x)) = x$. It follows that $\pi(t_0,x) = \pi'(k_0^{-1},x)$ for all $x\in M$. We have $t_0 = k_0^{-1}$ and t_0 = identity if and only if k_0 = identity. Hence A_p and S have a non-trivial element t_0 in common. By Lemma 2, A_p cannot act effectively on M. A contradiction! Consequently, $(S\times A_p,M,\pi'')$ must be an effective transformation group. There exists an orbit $\pi'(S,x_0)$ at some point $x_0\in M$, whose dimension is one. It follows that the dimension of the orbit $\pi''(S\times A_p,x_0)$ with $\pi'(S,x_0)$ as a subset, must be greater than or equal to one. Then, by a result of Bredon, (See [1]), which says that if a compact group acts effectively on a connected n-manifold with an orbit whose dimension is greater or equal to $(n-2)$ then the group must be Lie,

$K \times A_p$ is a Lie group. It follows that A_p is a Lie group. A contradiction! Hence the statement of the theorem is true.

<u>Remark</u>. In 1969, Peter Orlik and Frank Raymond (See [9]) classified all effective, circle transformation groups on closed, connected, 3-manifolds. These manifolds must be either (1) $S^3, S^2 \times S^1, N, P^2 \times S^1$, $L(p,q)$, where $L(p,q)$ is a lens space with $p > 0$, $q \geq 0$ and p and q are relatively prime, S^n is an n-sphere, P^2 is a projective plane and N is a non-trivial S^2 bundle over S^1, (2) a connected sum of the above spaces, (3) $K \times S^1$, KS, where K is a klein bottle and KS is a non-trivial klein bottle bundle over S^1, (4) a quotient of SO(3) or Sp(1) by a finite, non-abelian discrete subgroup or (5) an Eilenberg-MacLane space $K(\pi,1)$ whose fundamental group has infinite cyclic center.

<u>Theorem 3</u>. Let M be a closed, connected 2-manifold. If (Z,M,π) is an almost periodic, discrete flow, then it is either periodic or effective. If it is effective, then M must be S^2, P^2, KL or T^2, where T^2 is a 2-dim torus and the embedding problem of (Z,M,π) can be solved.

<u>Proof</u>. It is easy to see that if (Z,M,π) is not effective, it must be periodic. Let $Z = \{\phi^n | n = \text{integer}\}$. Define a map f from Z into M^M by $f(z) = \{z(x) | x \in M\}$, where $z \in Z$. Let $\overline{f(Z)} = G$. Then G is a compact topological group. If (Z,M,π) is effective, then, by Lemma 1, there is an effective transformation group (G,M,π'), with the compact group G as above, such that the following diagram is commutative:

$$\begin{array}{ccc} Z\quad M & \xrightarrow{\pi} & M \\ f\big\downarrow\ \ i\big\downarrow & & i\big\downarrow \\ G\quad M & \xrightarrow{\pi'} & M \end{array}$$

Since M is 2-manifold and (G,M,π) is effective, G must be a Lie group and the connected component G_0 of the identity is non-trivial. It follows that G_0 must be a torus. There exists an integer r such that $f(\phi^r) \in G_0$ and $f(\phi^r) \neq$ identity. Let $Z' = \{(\phi^r)^n | n = \text{integer}\}$. There exists a one-parameter subgroup R in G_0 such that $\pi(h,x) = \pi'(f(h),x)$ where $h \in Z'$ and $x \in M$. Hence the embedding problem (Z,M,π) is solved.

From known results in compact transformation groups, dim $(G_0) \leq \frac{1}{2} \ell(\ell+1)$ where ℓ is the max. dim. of all orbits in (G_0,M,π). It is not hard to see that ℓ must be one or two and M must be S^2, P^2, KL or T^2 (See [7]).

REFERENCES

[1] G.E. Bredon, "Some theorems on transformation groups", Annals of Mathematics 67 (1958), 104-118.

[2] Hsin Chu, "On the structure of almost periodic transformation groups", Pacific J. Math. 38 (1971), 359-364.

[3] W.H. Gottschalk and G.A. Hedlund, "Topological dynamics", Amer. Math. Soc. Colloq., Pub. 36(1955).

[4] K.H. Hofmann and P.S. Mostert, "Compact groups acting with (n-1)-dimensional orbits on subspaces of n-manifolds", Math. Ann. 167 (1966), 224-239.

[5] D. Montgomery, "Finite dimensionality of certain transformation groups", Illinois J. Math. 1(1957), 28-35.

[6] D. Montgomery and L. Zippin, Topological Transformation Groups, Interscience, New York, 1955.

[7] P.S. Mostert, "On a compact Lie group acting on a manifold", Ann. of Math. 65(1957), 447-455.

[8] W.D. Neumann, "3-dimensional G-manifolds with 2-dimensional orbits", Conference on Compact Transformation Groups, Edited by P.S. Mostert, Springer-Verlag, (1969), 220-222.

[9] P. Orlik and F. Raymond, "Actions of SO(2) on 3-manifolds", Conference on Compact Transformation Groups, Edited by P.S. Mostert, Springer-Verlag, (1969), 297-317.

[10] F. Raymond, "Cohomological and dimension theoretical properties of orbit spaces of p-adic actions", Conference on Compact Transformation Groups, Edited by P.S. Mostert, Springer-Verlag, (1969), 354-365.

[11] R.F. Williams, "Compact non-Lie groups", Conference on Compact Transformation Groups, Edited by P.S. Mostert, Springer-Verlag, (1969), 366-369.

[12] C.T. Yang, "p-adic transformation groups", Mich. Math. J. 7(1960), 201-218.

UNIVERSITY OF MARYLAND, COLLEGE PARK

A GROUP ASSOCIATED WITH AN EXTENSION

Robert Ellis(*)

An extension of a transformation group (Y,T) is a transformation group (X,T) together with a homomorphism φ of (X,T) onto (Y,T). The study of extensions is important because it has been recognized that many theorems have a so-called relativized version. Thus for example (X,T) itself may not be almost periodic but it may be almost periodic over (Y,T). Then information about (Y,T) may be "lifted" to (X,T). Another reason for studying extensions is that there are several structure theorems which tell how certain transformation groups are built up from "nice ones" by means of extensions. Again this leads one to inquire as to which properties of (Y,T) lift to (X,T).

In my investigations of the problems involved in the study of extensions I have repeatedly encountered a certain group whose role proved essential. I would like to describe this group and illustrate its use in the application of one structure theorem and the proof of another.

The study of extensions is facilitated by considering pointed transformation groups and requiring that φ map the base point of

(*). Research supported by NSF Grant GP-29321

X onto that of Y. Then φ becomes unique and may be omitted from the discussion. Moreover it is convenient to assume that $\mathcal{C}(Y) \subset \mathcal{C}(X)$, to identify X and Y with the maximal ideal spaces $|\mathcal{C}(X)|$ and $|\mathcal{C}(Y)|$ respectively, and to assume that φ is induced by the inclusion mapping of $\mathcal{C}(Y)$ into $\mathcal{C}(X)$. From this point of view $\mathcal{C}(X) \subset \mathcal{C}(M)$ for every pointed minimal transformation group X, where M is the universal minimal set. The minimal set M has a natural semigroup structure and its base point is chosen to be an idempotent u. Henceforth I shall be concerned only with T-subalgebras, $\mathfrak{A}$ of $\mathcal{C}(M)$ and their associated transformation groups $(|\mathfrak{A}|,T)$. (For details concerning this point of view see [1]).

It turns out that $G = Mu$ is a subgroup with identity u of the semigroup M and that to every T-subalgebra, $\mathfrak{A}$ may be associated a group $\mathfrak{G}(\mathfrak{A})$. The group G is just the group of automorphisms of M and $\mathfrak{G}(\mathfrak{A})$ may be thought of as the subgroup of G consisting of those automorphisms which reduce to the identity on $|\mathfrak{A}|$. Moreover each T-subalgebra, $\mathfrak{A}$ induces a topology $\tau(\mathfrak{A})$ on G. These topologies have the following properties: (i) $\tau(\mathfrak{A}) \subset \tau(\mathfrak{B})$ if $\mathfrak{A} \subset \mathfrak{B}$, (ii) $(G,\tau(\mathfrak{A}))$ is compact (but not Hausdorff), and (iii) $\mathfrak{G}(\mathfrak{A})$ is τ-closed, where $\tau = \tau(\mathcal{C}(M))$.

Now let $\mathfrak{F}$ and $\mathfrak{A}$ be T-subalgebras of $\mathcal{C}(M)$ with $\mathfrak{F} \subset \mathfrak{A}$. Let $F = \mathfrak{G}(\mathfrak{F})$, $A = \mathfrak{G}(\mathfrak{A})$, and let $\mathfrak{n}_{\mathfrak{A}}$ be the collection of $\tau(\mathfrak{A})$ neighborhoods of the identity u of G. The group associated with the extension of $|\mathfrak{F}|$ by $|\mathfrak{A}|$ is $H(F;\mathfrak{A}) = \text{cls}\{F \cap V \mid V \in \mathfrak{n}_{\mathfrak{A}}\}$ where "cls" denotes the closure with respect to the τ-topology. (This is not precisely the same as in [1] but the two have recently been shown to be equivalent). Then $H(F,\mathfrak{A})$ is actually a $\tau(\mathcal{A})$-closed subgroup which contains A. Indeed it may be characterized as the smallest subgroup K of F such that $(F|K, \tau(\mathcal{A}))$ is compact Hausdorff. If $\mathcal{A}$ is an almost periodic extension of $\mathcal{F}$, then $H(F,\mathcal{A}) = A$. However, the converse is not true without some additional assumption on the extension.

The definition of $H(F,\mathcal{A})$ makes sense even if $\mathcal{F}$ is not assumed to be a subalgebra of $\mathcal{A}$, and indeed later in this paper I will use $H(F,\mathcal{A})$ in such contexts.

The first situation to which I would like to apply the notions outlined above is that involving disjointness. Two transformation groups $\mathcal{A}$ and $\mathcal{B}$ are disjoint if their product $|\mathcal{A}| \times |\mathcal{B}|$ is minimal under the diagonal action; they have no common factor if $\mathcal{A} \cap \mathcal{B} = C$, the algebra of constant functions. It is easy to see that if $\mathcal{A}$ and $\mathcal{B}$ are disjoint then they have no common factor. For general T the converse is false but for abelian T the problem is still open. There are, however, some partial converses due to Peleg [3] and Keynes [2] which I would like to discuss from the above point of view.

Let T be abelian. Then $\mathcal{A}$ and $\mathcal{B}$ are disjoint if and only if $AB = G$ where $A = \mathfrak{G}(\mathcal{A})$ and $B = \mathfrak{G}(\mathcal{B})$. Although the above characterization does not hold in the non-abelian case it can be shown that even in that case the heart of the problem is to show that $AB = G$. The way this is done is to use a structure theorem and induction. Thus, for example suppose there exists a family $(\mathcal{A}_\alpha \mid \alpha \leq \nu)$ of transformation groups such that (i) $A_0 B = G$ and $A_\nu = A$, (ii) $\mathcal{A}_\alpha \subset \mathcal{A}_{\alpha+1}$ $(\alpha < \nu)$, (iii) $\mathcal{A}_\alpha = \overline{\cup\{\mathcal{A}_\beta \mid \beta < \alpha\}}$ for all limit ordinals $\alpha \leq \nu$, (iv) $A_{\alpha+1} = A_\alpha$ or $\mathcal{A}_\alpha \leq_\rho \mathcal{A}_{\alpha+1}$ (i.e. $\mathcal{A}_{\alpha+1}$ is an almost periodic extension of $\mathcal{A}_\alpha$) $(\alpha < \nu)$; where $A_\alpha = \mathfrak{G}(A_\alpha)$. Then in order to prove that $AB = G$ one needs to show (1) if $(A_\beta | \beta < \alpha)$ is a decreasing family of subgroups with intersection A_α , and if $A_\beta B = G$ $(\beta < \alpha)$, then $A_\alpha B = G$, and (2) if $A_\alpha B = G$ and $\mathcal{A}_\alpha \leq_\rho \mathcal{A}_{\alpha+1}$ then $A_{\alpha+1}B = G$. Condition (iii) states that $|\mathcal{A}_\alpha|$ is the inverse limit of the family, $(|\mathcal{A}_\beta| / \beta < \alpha)$ and (1) allows one to conclude that if $\mathcal{A}_\beta$ and $\mathcal{B}$ are disjoint $(\beta < \alpha)$, then $\mathcal{A}_\alpha$ and $\mathcal{B}$ are also disjoint. The proof of (1) is straightforward and uses nothing more than the compactness of the various groups involved.

The proof of (2) rests upon the following observations and results: (a) $\mathcal{A}_\alpha \leq_\rho \mathcal{A}_{\alpha+1} \cap (\mathcal{A}_\alpha \vee \mathcal{B})$, (b) $\mathfrak{G}(\mathcal{A}_{\alpha+1} \cap (\mathcal{A}_\alpha \vee \mathcal{B})) = (A_\alpha \cap B)\, A_{\alpha+1}$, (c) $H(A_\alpha, \mathcal{B}) = H(A_\alpha, \mathcal{A}_\alpha \vee \mathcal{B})$ (d) $H(A_\alpha, \mathcal{A}_{\alpha+1} \cap (\mathcal{A}_\alpha \vee \mathcal{B})) = (A_\alpha \cap B) A_{\alpha+1}$

and

THEOREM 1: <u>Let</u> $\mathcal{F} \subset \mathcal{A} \cap \mathcal{B}$, $V_0 \in \mathfrak{n}_{\mathcal{B}}$, <u>and</u> $BA \supset H(F, \mathcal{B}) \cup V_0$. Then $H(F, \mathcal{B}) \subset B\, H(A, \mathcal{B})$.

To apply theorem 1 to the present situation set $\mathcal{F} = C$, $\mathcal{B} = \mathcal{B}$, and $\mathcal{A} = \mathcal{A}_\alpha$. Then since $B\, A_\alpha = G$ we may conclude that

$H(G, \mathcal{B}) \subset B\, H(A_\alpha, \mathcal{B}) = B\, H(A_\alpha, \mathcal{A}_\alpha \vee \mathcal{B}) \subset B\, H(A_\alpha, \mathcal{A}_{\alpha+1} \cap (\mathcal{A}_\alpha \vee \mathcal{B})) =$
$B\, (A_\alpha \cap B)\, A_{\alpha+1} = B\, A_{\alpha+1}$.

Now $\mathcal{A} \cap \mathcal{B} = C$ implies that $\mathcal{A} \cap (\mathcal{B} \cap \mathcal{E}) = C$, where $\mathcal{E}$ is the algebra of all almost periodic functions. (Thus $|\mathcal{B} \cap \mathcal{E}|$ is the structure group of $|\mathcal{B}|$) . Then $B\, E\, A = \mathfrak{G}(\mathcal{A} \cap \mathcal{B} \cap \mathcal{E}) = G$. Moreover in this case $E \subset H(G, \mathcal{B})$. Since $A \subset A_{\alpha+1}$, $G = B\, E\, A \subset$
$B\, H(G, \mathcal{B})\, A_{\alpha+1} \subset B\, B\, A_{\alpha+1} = B\, A_{\alpha+1}$.

Before leaving this topic I would like to make a few remarks about conditions (i), (ii), (iii) and (iv) above.

(1) The various structure theorems are of this form; i.e. conditions are imposed on $\mathcal{A}$ and a family $(\mathcal{A}_\alpha \mid \alpha \leq \nu)$ constructed satisfying conditions similar to (i), (ii), (iii) and (iv) .

(2) The avove assumptions on the family $(\mathcal{A}_\alpha \mid \alpha \leq \nu)$ are weaker than the ones satisfied by families constructed in the Furstenberg or the Veech structure theorems. Thus the above results apply whenever $\mathcal{A}$ satisfies the assumptions of one of these structure theorems.

In the preceding discussion I was concerned primarily with the case $H(F,\mathcal{A}) = A$. Now I would like to discuss the other extreme, i.e. $H(F,\mathcal{A}) = F$. These considerations arose in my work on the Veech structure theorem. Recently Veech [4] proved: THEOREM 2. Let $|\mathcal{A}|$ be a metrizable transformation group with a residual set of distal points. Then there exists a countable ordinal ν and a family $(\mathcal{A}_\alpha \mid \alpha \leq \nu)$ of transformation groups such that (i) $\mathcal{A}_0 = C$, (ii) $|\mathcal{A}_\alpha| = \text{inv lim } (|\mathcal{A}_\beta| \mid \beta < \alpha)$ if α is a limit ordinal $\leq \nu$, (iii) $\mathcal{A}_{\alpha+1}$ is either an almost automorphic or an almost periodic extension of $\mathcal{A}_\alpha$ $(\alpha < \nu)$. (iv) $\mathcal{A}_\nu$ is an almost automorphic extension of $\mathcal{A}$. To explain the terms involved above let me introduce some notation. Let $\mathcal{F} \subset \mathcal{A}$, $x \in |\mathcal{A}|$ then $x|\mathcal{F}$ will denote the image of x in $|\mathcal{F}|$; $R(\mathcal{A}: \mathcal{F}) = \{(x,y) \mid x,y \in |\mathcal{A}|,\ x|\mathcal{F} = y|\mathcal{F}\}$, and $P(\mathcal{A}: \mathcal{F}) = \cap \{\eta T \mid \eta \text{ index on } |\mathcal{A}|\} \cap R(\mathcal{A}: \mathcal{F})$. Then $x \in |\mathcal{A}|$ is a distal point if $xP(\mathcal{A}: C) = x$ and $\mathcal{A}$ is an almost automorphic extension of $\mathcal{F}$ if there exists $x \in |\mathcal{A}|$ with $x R(\mathcal{A}: \mathcal{F}) = x$.

By analyzing the groups of the various extensions involved in this situation I was able to prove the Veech structure theorem assuming only that the set of distal points was not empty. As Veech pointed out this implies that if a metrizable transformation group has one distal point, then the distal points form a residual set.

Let me now outline what is involved in the proof. As with most structure theorems of the above type the main problem is to insert a non-trivial almost periodic extension $\mathcal{L}$ between $\mathcal{F}$ and $\mathcal{A}$. Here $|\mathcal{A}|$ is assumed to be metrizable and to have a distal point which may be assumed to be the base point, u . By an ingenious argument Veech shows that by taking almost automorphic extensions of $\mathcal{F}$ and $\mathcal{A}$ if necessary we may assume that the map $|\mathcal{A}| \to |\mathcal{F}|$ is open.

Now the obvious candidate for $\mathcal{L}$ is just $\mathcal{A} \cap \mathcal{F}^{\#}$ where $\mathcal{F}^{\#}$ is the set of all almost periodic functions over $|\mathcal{F}|$ (see [1]) . Then $\mathcal{F} \neq \mathcal{F}^{\#} \cap \mathcal{A}$ if and only if $F \neq \mathfrak{G}(\mathcal{F}^{\#} \cap \mathcal{A})$. In this connection the following lemmas are relevant. # LEMMA 1. *Let* $|\mathcal{A}| \to |\mathcal{F}|$ *be open and let* $|\mathcal{A}|$ *have a distal point. Then the almost periodic points of* $R(\mathcal{A}\colon \mathcal{F})$ *are dense in* $R(\mathcal{A}\colon \mathcal{F})$.

LEMMA 2. *Let the almost periodic points of* $R(\mathcal{A}\colon \mathcal{F})$ *be dense in* $R(\mathcal{A}\colon \mathcal{F})$. *Then* $H(F,\mathcal{A}) = \mathfrak{G}(\mathcal{F}^{\#} \cap \mathcal{A})$. # From lemmas 1 and 2 we conclude that a non-trivial almost periodic extension of $\mathcal{F}$ may be inserted between $\mathcal{F}$ and $\mathcal{A}$ if $H(F,\mathcal{A}) \neq F$. This leads us to examine the implications of the relation $H(F,\mathcal{A}) = F$. Here the following general result is relevant.

PROPOSITION 1. *Let* $H(F,\mathcal{A}) = F$ *and let* $P(\mathcal{A}\colon \mathcal{F}) = \cap \{\eta_i T \cap R(\mathcal{A}\colon \mathcal{F}) \mid i \in I\}$ *where* $(\eta_i \mid i \in I)$ *is a countable set of indices of* $|\mathcal{A}|$. *Then there exists an idempotent* w *in* M *such that* $w|\mathcal{F} = u|\mathcal{F}$ *and the set* $L = \{p \mid p \in \overline{Fw}\, , (p|\mathcal{A}, w|\mathcal{A}) \in P(\mathcal{A}\colon \mathcal{F})\}$ *is a residual subset of* $\overline{Fw}$. (here $\overline{Fw}$ denotes the closure of Fw with respect to the usual topology on M) # Thus in the case under consideration if $H(F,\mathcal{A}) = F$ we may draw the following conclusions: (1) $w|\mathcal{A} = u|\mathcal{A}$ since $w|\mathcal{F} = u|\mathcal{F}$ and $w^2 = w$ implies that $(w|\mathcal{A}, u|\mathcal{A}) \in P(\mathcal{A}\colon \mathcal{F})$ (recall that $u|\mathcal{A}$ is a distal point) (2) If $p \in L$, then $p|\mathcal{A} = u|\mathcal{A}$ (3) $F = A$; this follows from (2) and the fact that $F \subset L$. Consequently $\mathcal{F}$ is a proximal extension of $\mathcal{A}$ with a distal point $x_0 = u|\mathcal{A}$. Hence the fiber $x_0 R(\mathcal{A}\colon \mathcal{F})$ over $u|\mathcal{F}$

consists of the singleton x_0 . Since the map $|\mathcal{A}| \to |\mathcal{F}|$ is open, this implies $\mathcal{F} = \mathcal{A}$.

The above proof indicates that the assumption that $|\mathcal{A}|$ be metirzable can be replaced by the assumption that $|\mathcal{A}|$ be quasi-separable. (A transformation group is quasi-separable if it is the inverse limit of metrizable transformation groups. Thus if T is countable every compact minimal set is quasi-separable)

THEOREM 3. *Let* $|\mathcal{A}|$ *be a point-distal quasi-separable transformation group. Then there exists a family of* T-*subalgebras* $(\mathcal{A}_\alpha | \alpha \leq \nu)$ *such that:* (i) $\mathcal{A}_0 = C$, (ii) $|\mathcal{A}_\alpha| = \text{inv lim } (|\mathcal{A}_\beta| \, |\beta < \alpha)$ *if* α *is a limit ordinal* $\leq \nu$ (iii) $\mathcal{A}_{\alpha+1}$ *is either a proximal or an almost periodic extension of* $\mathcal{A}_\alpha$ $(\alpha < \nu)$. (iv) $\mathcal{A}_\nu$ *is a proximal extension of* $\mathcal{A}$.

The difference in the conclusions of theorem 3 from those of theorem 2 is that in the former ν need not be a countable ordinal and $\mathcal{A}_{\alpha+1}$ may be a proximal rather than an almost automorphic extension of $\mathcal{A}_\alpha$. Of course in this case (when $|\mathcal{A}|$ is not assumed to be. metrizable) we cannot conclude that the set of distal points is residual. As with most theorems of this type, there are relativized versions of theorems 2 and 3 .

Detailed proofs of the results mentioned in this paper will appear elsewhere.

REFERENCES

1. R. Ellis, Lectures on Topological Dynamics, W.A. Benjamin, New York, 1969

2. Harvey B. Keynes, Disjointness in transformation groups

3. R. Peleg, Weak disjointness of transformation groups P.A.M.S. (to appear)

4. W.A. Veech, Point-distal flows, Amer. Jour. Math. 92 (1970), 205-242.

University of Minnesota

THE UNIQUE ERGODICITY OF THE HOROCYCLE FLOW

HARRY FURSTENBERG

Introduction. Professor Hedlund's name is associated with two of the classical dynamical systems of topological dynamics, the geodesic flow and the horocycle flow on surfaces of constant negative curvature ([4], [5], [6]). Both flows arise in a similar way. Geometrically, they both represent motion of unit tangent vectors along lines of constant geodesic curvature on compact surfaces of constant negative curvature. The geodesics have 0 geodesic curvature and the horocycles have maximal constant curvature without being closed. Both have similar group-theoretic descriptions: both are actions of non-compact one-parameter subgroups of $G = SL(2, \mathbb{R})$ on a compact quotient of G by a discrete subgroup Γ. The principal results for these flows were discovered in the 30's; namely the ergodicity of the geodesic flow (which implies that "almost all" orbits are dense) and the minimality of the horocycle flow (which is equivalent to the assertion that all orbits are dense) ([4], [5], [6], [7]). In the meantime, however, the two flows have fared differently. The geodesic flow is very well understood and has become the prototype of a large class of flows (Anosov flows), whereas the horocycle flow seems to stand in class by itself. Even confining ones attention to homogeneous flows, the phenomena associated with the geodesic flow, namely, ergodicity and mixing, are better understood than that associated with the horocycle flow, namely, minimality. To make this contention precise, let G be a simple Lie group and Γ a uniform subgroup (discrete and cocompact) and let T be a one-parameter subgroup of G. A reasonable condition on T (and in fact a necessary condition) for the ergodicity of the action of T on G/Γ turns out to be sufficient (T not have compact closure in G; see [8]). Now an obvious necessary condition for the minimality of the action of T on G/Γ is that there are no periodic cycles ($\Longleftrightarrow gTg^{-1} \cap \Gamma = \{e\}$, $g \in G$). Is this condition sufficient? So far one knows the answer only for $G = SL(2,\mathbb{R})$ when under these hypotheses the flow is the horocycle flow and the result is Hedlund's theorem. Another more general question is the following. Is every minimal homogeneous flow on a compact homogeneous space of the form G/Γ also uniquely ergodic? Here G is a Lie group and Γ is a uniform subgroup so that there is a unique probability measure $m_{G/\Gamma}$ on G/Γ preserved by G. The unique ergodicity of the action of a subgroup T on G/Γ means that $m_{G/\Gamma}$ is the only probability measure on G/Γ preserved by the action of T. When T is abelian this implies that T is minimal on G/Γ (because T then leaves invariant some probability measure on any closed T-invariant subset) and unique ergodicity implies both ergodicity and minimality. The example of nilflows which have been thoroughly studied bears out the conjecture that minimality does imply unique ergodicity for homogeneous flows. (See [2] for a counterexample in the general case.) In this paper we show that the horocycle flow also supports this conjecture. But the general question remains open.

We hope that the method developed here enables one to make headway with some of these questions. While we shall deal here exclusively with the classical horocycle flow, a similar method seems to yield at least the following: if $T \subset SL(m, \mathbb{R})$ is the subgroup leaving pointwise fixed an r-dimensional subspace of $\mathbb{R}^m$, where $2r < m$, then T acts uniquely ergodically on $SL(m, \mathbb{R})/\Gamma$. For $m > 2$ this overlaps the minimality result of L. Greenberg in [1, appendix].

1. Measures on Locally Compact Spaces. We shall be discussing borel measures on a locally compact, σ- compact, metrizable space X, by which we shall mean σ- additive measures taking values in $[0,\infty]$ for all borel subsets of X, but finite for compact sets. We denote the space of these by $\mathcal{M}(X)$. $\mathcal{M}(X)$ may be identified with the space of positive linear functionals on the vector space $\mathcal{C}_0(X)$ of continuous functions having compact support in X. If g is a borel measurable map from one space X to another Y, we define $g: \mathcal{M}(X) \longrightarrow \mathcal{M}(Y)$ by $g\mu(A) = \mu(g^{-1}A)$, $A \subset Y$. If G is a group of such transformations of X to itself, we say a measure $\mu \in \mathcal{M}(X)$ is G-invariant if $g\mu = \mu$ for all $g \in G$. We denote by $\mu_1 \prec \mu_2$ the absolute continuity of μ_1 with respect to μ_2, and we say μ_1 is equivalent to μ_2 if $\mu_1 \prec \mu_2$ and $\mu_2 \prec \mu_1$. Note that under our hypotheses on X, every measure is equivalent to a finite measure (i.e., a measure μ with $\mu(X) < \infty$).

The following proposition will be indispensable.

Proposition 1.1. If $Z = X \times Y$ where X, Y are locally compact, σ- compact, and metrizable, then every measure $\mu \in \mathcal{M}(Z)$ admits a decomposition $d\mu(x,y) = d\lambda_y(x)\,d\nu(y)$ in the sense that

$$\int_Z f(x,y)\,d\mu(x,y) = \int_Y \{ \int_X (x,y)\,d\lambda_y(x) \}\,d\nu(y)$$

for $f \in \mathcal{C}_0(Z)$. Here λ_y is a borel measurable function from Y to $\mathcal{M}(X)$ and $v \in \mathcal{M}(Y)$. This decomposition is unique in the sense that if $d\lambda'_y(x)\,d\nu'(y)$ is another such decomposition then ν and ν' are equivalent and $\lambda'_y = \frac{d\nu}{d\nu'}(y)\,\lambda_y$ for almost all y (relative to ν). Finally, if μ is a finite measure on Z then the measure ν can be taken as the projection of μ on Y. In particular, the equivalence class of ν is determined by that of μ.

We shall sketch a proof of this proposition which may be considered a "folk theorem". One sees readily that it suffices to prove the statements of the proposition in the case that μ is a finite, or even probability, measure. In this case we choose an appropriate countable dense subset of $\mathcal{C}_o(X)$, say, $\{f_n(x)\}$. We let P denote the projection from $L^2(Z, \mu)$ to $L^2(Y, \nu)$ where $\nu \in \mathcal{M}(Y)$ is the image of $\mu \in \mathcal{M}(Z)$, so that $L^2(Y, \nu)$ may be regarded as a closed subspace of $L^2(Z, \mu)$, then $(Pf_n)(y)$ is defined almost everywhere in Y and fixing $y \in Y$ one finds that for almost all y, $Pf_n(y)$ extends to a positive linear functional on $\mathcal{C}_o(X)$. This yields a measurable map $Y \longrightarrow \mathcal{M}(X)$ and it is not difficult to show that this map $y \longrightarrow \lambda_y$ together with ν yield a decomposition of μ as desired. One should observe that the operator P here is essentially the <u>conditional expectation</u> for the borel field of Z with respect to the subfield defined by sets of Y. For more details we refer the reader to [2].

One application of this proposition is to homogeneous spaces of Lie groups. Recall that a locally compact group is called <u>unimodular</u> if there exists a measure on the group which is invariant under both left and right translations.

<u>Proposition 1.2.</u> Let H be a closed unimodular subgroup of the Lie group G. There is a one-one correspondence between $\mathcal{M}(G/H)$ and measures on G invariant with respect to the action of right translation by elements of H. The correspondence is given by $\omega \longleftrightarrow \pi$ where

$$\int_G f(g)\, d\,\omega(g) = \int_{G/H} \int_H f(gh)\, dh\, d\pi(gH).$$

<u>Proof.</u> It is clear that the integral to the right defines an H-invariant measure on G. We have to show that every such measure arises in this way. Under our hypotheses it is known that there exists a borel measurable crossection $\sigma : G/H \longrightarrow G$ where we re-

gard G as a fibre bundle over G/H with fibre H. So, for measure-theoretic purposes we may consider G as a product $G/H \times H$ where $g \longleftrightarrow (x, h)$ with $x = gH$ and $g = \sigma(gH)h$. We then write $d\omega(g) = d\omega(x, h) = d\lambda_x(h)d\pi'(x)$. Since ω is right H-invariant, we have (by the uniqueness in Proposition 1.1) that for almost all x, λ_x is right invariant on H. (This argument uses the fact that the measure ν in Proposition 1.1 can obviously be chosen arbitrarily in its equivalence class, and so if ω and a translate of ω by $h \in H$ are identical, both may be represented with the same $\pi' \in \mathcal{M}(G/H)$). Since H is unimodular we have $d\lambda_x(h) = \varphi(x)dh$, dh being haar measure on H, and so, setting $d\pi(x) = \varphi(x)d\pi'(x)$ we obtain the desired result. Once again the uniqueness in Proposition 1.1 shows that π is uniquely determined by ω once the haar measure of H is fixed.

From this one deduces the following.

Proposition 1.3. If H, L are both closed unimodular subgroups of the Lie group G, there is a one-one correspondence between L-invariant measures on G/H and measures on G that are invariant under left translations by elements of L and right translations by elements of H.

Finally we have

Theorem 1.4. If H, L are both closed unimodular subgroups of the Lie group G, there is a one-one correspondence between L-invariant measures on G/H and H-invariant measures on G/L.

For the proof one lifts measures of both types to the group G using the foregoing proposition, and associates to a measure $\omega \in \mathcal{M}(G)$ the measure ω^* defined by $\omega^*(A) = \omega(A^{-1})$.

2. Homogeneous spaces of $SL(2, \mathbb{R})$. For the remainder of this paper G will denote the 2×2 unimodular group $SL(2, \mathbb{R})$. Γ will denote a fixed discrete subgroup of G having compact quotient G/Γ. In this section we shall discuss some other homogeneous spaces of G which play an important role in our subsequent discussion. We denote by K, A, N, respectively, the subgroups of matrices of the form

$$k_\theta = \begin{pmatrix} \cos\theta & \sin\theta \\ -\sin\theta & \cos\theta \end{pmatrix} \quad a_s = \begin{pmatrix} s & o \\ o & s^{-1} \end{pmatrix}, \; s > o \quad n_t = \begin{pmatrix} 1 & t \\ o & 1 \end{pmatrix}.$$

the horocycle flow is the flow on G/Γ determined by the subgroup N : $\tau_t(g\Gamma) = n_t g\Gamma$. The center of G consists of $\{\pm 1\}$ and we denote by T the subgroup of upper triangular matrices: $T = \{\pm 1\}\cdot A\cdot N$.

With the natural action of G on the vector space $\mathbb{R}^2$, we find that $\mathbb{R}^2 - \{0\}$ is a homogeneous G-space which may be identified with G/N. G also acts on the complex upper half plane $\mathcal{H} = \{x+iy \mid y>0\}$ by

$$\begin{pmatrix} a & b \\ c & d \end{pmatrix}(w) = \frac{aw+b}{cw+d}.$$

This action is transitive and we recognize the stability group of $i \in \mathcal{H}$ to be just K, whence we may identify G/K with $\mathcal{H}$. It will also be useful to identify $\mathcal{H}$ with the unit disc $\mathcal{D}$ by

$$z = \sigma(w) = \frac{w-i}{w+i}.$$

This determines an action of G on $\mathcal{D}$ which we again denote $z \longrightarrow g(z)$, but with a different formula than that given above. We frequently switch between the three spaces G/K, $\mathcal{H}$, and $\mathcal{D}$, using whichever suits our purposes best.

Since T is the subgroup of G taking the X-axis in $\mathbb{R}^2$ to itself, we may identify G/T by the set of lines through the origin of $\mathbb{R}^2$; i.e., $G/T \cong P'$, the projective line. Note that it also corresponds to the boundaries $\partial\mathcal{D}$ and $\mathbb{R} \cup \{\infty\}$ of $\mathcal{D}$ and $\mathcal{H}$ respectively. If $x \in \mathbb{R}^2 - \{0\}$ we denote by $\langle x \rangle$ the point of P' corresponding to the line in $\mathbb{R}^2$ joining x to the origin. One

One may then identify G/K, P', $\mathbb{R} \cup \{\infty\}$ and $\partial \mathcal{D}$ as follows

$$(2.1) \qquad \begin{pmatrix} a & b \\ c & d \end{pmatrix} T \longleftrightarrow \left\langle \begin{pmatrix} a \\ c \end{pmatrix} \right\rangle \longleftrightarrow \frac{c}{a} \longleftrightarrow \frac{c - ai}{c + ai} .$$

Note that G preserves lebesgue measure on $\mathbb{R}^2$ and hence on $\mathbb{R}^2 - \{0\} = G/N$. The unique ergodicity of the horocycle flow is the statement that N preserves a unique probability measure on G/Γ. Since N and Γ are unimodular we may apply Theorem 1.4 to obtain

<u>Theorem 2.1</u>. The horocycle flow is uniquely ergodic iff the only Γ- invariant measures on $\mathbb{R}^2 - \{0\}$ are the constant multiples of lebesgue measure.

<u>3. Positive Harmonic Functions on G/K</u>. We shall consider harmonic functions on $\mathcal{D}$ and $\mathcal{H}$, or, equivalently, on G/K. Note that a function on the coset space G/K can be thought of as a function $f(g)$ on G satisfying $f(gk) = f(g)$ for $k \in K$.

The following is basic for our considerations.

<u>Proposition 3.1</u>

(a) the functions

$$P(\zeta, z) = \frac{1 - |z|^2}{|\zeta - z|^2},$$

$\zeta \in \partial \mathcal{D}$, $z \in \mathcal{D}$, are harmonic functions in $\mathcal{D}$ for each ζ.

(b) If $h(z)$ is a positive harmonic function in $\mathcal{D}$ it may be represented uniquely in the form

$$h(z) = \int_{\partial \mathcal{D}} P(\zeta, z)\, d\nu(\zeta)$$

where ν is a non-negative measure (clearly finite) on $\partial \mathcal{D}$.

(c) The integrals

$$I_p(r) = \int_0^{2\pi} |h(re^{i\theta})|^p d\theta \quad 1 < p < \infty$$

increase with r for any harmonic function $h(z)$ on $\mathcal{D}$. A necessary and sufficient condition for the measure ν in (b) to be of the form $d\nu(\zeta) = \psi(\zeta)dm(\zeta)$, $\psi \in L^p(\partial\mathcal{D}, m)$, m lebesgue measure on $\partial\mathcal{D}$, is that $I_p(r)$ be bounded as $r \nearrow 1$. In this case we say $h(z)$ is L^p- bounded.

(d) For $x \in \mathbb{R}^2 - \{0\}$, let $\|x\|$ denote the euclidean norm. Let ζ_x correspond to $\langle x\rangle$ in the identifications (2.1). Then we can write

$$(3.1) \qquad P(\zeta_x, g(0)) = \|x\|^2/\|g^{-1}x\|^2, \quad g(0) \in \mathcal{D}.$$

<u>Proof:</u> (a), (b), and (c) are classical (see, e.g., [9]). The identity in (d) can be (presumably) verified by direct computation, but we present a more transparent proof. If we write the Möbius transformation g of $\mathcal{D}$ as

$$g(z) = \frac{\alpha z + \beta}{\bar{\beta} z + \bar{\alpha}}, \quad |\alpha|^2 - |\beta|^2 = 1$$

we find that

$$|g^{-1'}(\zeta)| = \frac{1}{|\alpha - \beta\zeta|^2} = \frac{|1 - g(o)|^2}{|\zeta - g(o)|^2} = P(\zeta, z)$$

with $g(0) = z$. Now if m denotes lebesgue measure on $\partial\mathcal{D}$ we can write

$$|g^{-1'}(\zeta)| = \frac{dgm}{dm}(\zeta),$$

so that the Poisson kernel is represented as a Rodon-Nikodym derivative. To transfer this from $\partial\mathcal{D}$ to P', we must replace m by the K-invariant probability measure on P'. Denoting this measure by μ, we may write

$$P(\zeta_x, g(0)) = \frac{dg\mu}{d\mu}(\langle x\rangle).$$

Finally a simple geometric argument will convince the reader that

$$\frac{dg\mu}{d\mu}(\langle x \rangle) = \frac{\|x\|^2}{\|g^{-1}x\|^2}.$$

(See [3] for a general framework for this type of formula.) This proves the proposition.

We remark that the form of the function of g occurring in (3.1) is clearly dependent only on the coset gK.

The proposition shows that $\|g^{-1}x\|^{-2}$ defines a harmonic function on G/K. More generally, if ν is a measure with compact support on $\mathbb{R}^2 - \{0\}$, the integral

$$\int \|g^{-1}x\|^{-2} d\nu(x) \tag{3.2}$$

defines a positive harmonic function on G/K. Furthermore in view of (b) and (d) of the proposition, we find that every positive harmonic function on G/K has such a representation (say, with ν concentrated on the unit circle of $\mathbb{R}^2$). Without any restriction on ν, this representation will not be unique, since for $x_2 = \lambda x_1$, the functions $\|g^{-1}x_1\|^{-2}$ and $\|g^{-1}x_2\|^{-2}$ are proportional.

To determine the extent to which a representation of a positive harmonic function in the form (3.2) is unique, we use the fact that $\mathbb{R}^2 - \{0\}$ is a fibre bundle over P' with fibre $\mathbb{R} - \{0\}$. With this structure each finite measure $\nu \in \mathcal{M}(\mathbb{R}^2 - \{0\})$ projects down to a measure $\bar{\nu} \in \mathcal{M}(P')$. Inasmuch as we can regard $\mathbb{R}^2 - \{0\}$ as the product, in a borel measurable fashion, of P' and $\mathbb{R} - \{0\}$, we may apply Proposition 1.1 to obtain

$$d\nu(x) = d\lambda_{\langle x \rangle}(t)\, d\bar{\nu}(\langle x \rangle). \tag{3.3}$$

(Here $t \in \mathbb{R} - \{0\}$. The measures $\lambda_{\langle x \rangle}$ are not uniquely determined by ν but depend also on the choice of a borel-measurable crossection $P' \longrightarrow \mathbb{R}^2 - \{0\}$.) We now find in evaluating the integral (3.2) in accordance with (3.3) and comparing the result with the representation of harmonic functions in Proposition 3.1 (b):

Theorem 3.2. Every measure ν of compact support on $\mathbb{R}^2 - \{0\}$ defines a positive harmonic function by

$$h(gK) = \int \|g^{-1}x\|^{-2} d\nu(x). \tag{3.2 bis}$$

Every positive harmonic function admits such a representation and for given $h(gK)$ the projection of the measure ν on P^1 is determined up to equivalence. This class is that of boundary measure occurring in the representation of Proposition 3.1.

4. Sketch of the Proof of Unique Ergodicity. According to Theorem 2.1 what must be shown is that if π is a Γ- invariant measure on $\mathbb{R}^2 - \{0\}$, then π is proportional to m, where m will heretofore denote lebesgue measure on $\mathbb{R}^2 - \{0\}$. We write $d\pi(x) = d\lambda_\xi(t)\, d\tilde{\pi}(\xi)$ in accordance with Proposition 1.1. Our first step will be to show that without loss of generality we may suppose all the measures λ_ξ absolutely continuous. We next show that it suffices to prove that π is absolutely continuous with respect to m. Decomposing m in the same way we find it sufficient to prove that $\tilde{\pi} \prec \tilde{m}$, the latter being, of course, lebesgue measure on P^1. We then show that the class of $\tilde{\pi}$ is determined by projecting onto P^1 the restriction π_Q of π to a sufficiently large compact set $Q \subset \mathbb{R}^2 - \{0\}$. Since this projection occurs in Theorem 3.2 in connection with the harmonic function

$$h_Q(gK) = \int_Q \|g^{-1}x\|^{-2} d\pi(x)$$

we find that we are called upon to study such functions. In particular it will follow from Proposition 3.1(c) that the measure class $\tilde{\pi}$ will be absolutely continuous iff $h_Q(gK)$ is L^2- bounded. The reason for choosing $p = 2$ will be apparent. (Once the result has been proven it will follow that $\pi = cm$ and h_Q is in fact bounded!)

It will now develop that to determine if $h_Q(gK)$ is L^2- bounded it will suffice to examine $h_Q(\gamma K)$ for $\gamma \in \Gamma$. Here we apply the Γ-invariance of to write

(4.1) $$h_Q(\gamma K) = \int_{\gamma^{-1}Q} \|x\|^{-2} d\pi(x).$$

The crux of our argument will be that for evaluating averages of expressions of the form (4.1) we can replace the measure π by m. Here the result will have to be true since m defines bounded harmonic functions. The details follow.

5. Preliminaries. We begin with

Lemma 5.1. If π is a Γ- invariant measure on $\mathbb{R}^2 - \{0\}$ and π is absolutely continuous (with respect to lebesgue measure), then π is proportional to lebesgue measure.

This amounts to the ergodicity of the action of Γ on $\mathbb{R}^2 - \{0\}$ which by the duality principle in [8] is equivalent to the ergodicity flow. (See [1] or [8].)

Lemma 5.1 reduces our problem to studying the equivalence class of the invariant measure π. We wish to reduce this further to the determination of the class of the "angular" part of the measure π.

Lemma 5.2. For $\ell > 0$ let $\ell\pi$ denote the measure $\ell\pi(A) = \pi(\ell^{-1}A)$. If $\psi \in C_o(0,\infty))$ we may form the measure $\pi_\psi = \int_0^\infty \ell\pi\, \psi(\ell)\,d\ell$. If each measure π_ψ is proportional to lebesgue measure on $\mathbb{R}^2 - \{0\}$ then so is π.

Proof: For we can choose $\psi_n \in C_o((0,\infty))$ with $\pi_{\psi_n} \longrightarrow \pi$ weakly in $\mathcal{M}(\mathbb{R}^2 - \{0\})$.

We can now enunciate

Proposition 5.3. For the proof of the unique ergodicity of the horocycle flow it suffices to show that if $\pi \in \mathcal{M}(\mathbb{R}^2 - \{0\})$ is Γ-invariant and π has a decomposition

$$d\pi(x) = d\lambda_{<x>}(t)\, d\tilde{\pi}\ <x>)$$

then $\tilde{\pi}$ is absolutely continuous on P^1.

Proof: For it suffices to show that π is absolutly continuous. Now by Lemma 5.2, it suffices to restrict our attention to measures of the form $(\pi')_\psi$. But if $\pi = (\pi')_\psi$, the measures $\lambda_{<x>}(t)$ are clearly absolutely continuous. Thus it suffices to show that the "angular" part, $\tilde{\pi}$ is absolutely continuous.

Lemma 5.4. There exists a compact set Q in $\mathbb{R}^2 - \{0\}$ with $\Gamma Q = \mathbb{R}^2 - \{0\}$.

This is clear from the cocompactness of Γ in G.

Now let π_Q denote the restriction of π to Q. We then have

Lemma 5.5. If π is a Γ-invariant measure in $\mathbb{R}^2 - \{0\}$, $\tilde{\pi}$ will be absolutely continuous iff the projection $\bar{\pi}_Q$ of π_Q on P^1 is absolutely continuous.

Proof: One direction is clear since $\pi_Q < \pi$. For the other direction assume that $\bar{\pi}_Q$ is absolutely continuous. Then so is $\gamma\bar{\pi}_Q = \bar{\pi}_{\gamma Q}$ for $\gamma \in \Gamma$. Now if a set in $\mathbb{R}^2 - \{0\}$ is a null set for all $\pi_{\gamma Q}$ then it must be a null set for π, since $\Gamma Q = \mathbb{R}^2 - \{0\}$. It follows that if a set is a null set for all $\bar{\pi}_{\gamma Q}$, it is also a null set for $\tilde{\pi}$. The result now follows.

Combining the last lemma with Proposition 5.3, Theorem 3.2, and Proposition 3.1(c) we find

Theorem 5.6. To prove the unique ergodicity of the horocycle flow it suffices to prove that the harmonic function

$$h_Q(gK) = \int_Q \|g^{-1}x\|^{-2} d\pi(x)$$

is L^2-bounded whenever π is a Γ-invariant measure.

6. L^2-boundedness of the Functions h_Q. Let π be a Γ-invariant measure on $\mathbb{R}^2 - \{0\}$.

Lemma 6.1. There exists a (finite) measure μ on G with compact support such that for each $\varphi \in C_0(\mathbb{R}^2 - \{0\})$

$$\int \varphi(x)\,dm(x) = \iint \varphi(gx)\,d\pi(x)\,d\mu(g).$$

Proof: Let $\tilde{\mu}$ be the G-invariant measure on G/Γ. Map G/Γ to G by some measurable crossection σ and let $\mu = \sigma(\tilde{\mu})$. Let π' be defined by

$$\int \varphi\,d\pi' = \iint \varphi(gx)\,d\pi(x)\,d\mu(g).$$

We claim π' is G-invariant. This will imply that π' is proportional to m and this yields the lemma. Fix $g_0 \in G$. Since $g_0\tilde{\mu} = \tilde{\mu}$ it follows that $g_0\mu$ and μ project onto the same measure on G/Γ. It is not hard to see that this means that we can decompose μ as $\mu = \Sigma\,\mu_\gamma = \Sigma\,\mu'_\gamma$ where each part μ_γ satisfies $g_0\mu_\gamma = \mu'_\gamma\gamma$, the right translate of μ'_γ by γ. Then

$$\begin{aligned}
&\int \varphi(g_0x)\,d\pi'(x) = \iint \varphi(g_0gx)\,d\pi(x)\,d\mu(g) = \\
&\sum_\Gamma \iint \varphi(g_0gx)\,d\pi(x)\,d\mu_\gamma(g) = \sum_\Gamma \iint \varphi(gx)\,d\pi(x)\,dg_0\mu_\gamma(g) = \\
&\sum_\Gamma \iint \varphi(gx)\,d\pi(x)\,d\mu' \quad (g) = \sum_\Gamma \iint \varphi(g\gamma\,x)\,d\pi(x)\,d\mu'_\gamma(g) = \\
&\sum_\Gamma \iint \varphi(gx)\,d_\gamma\pi(x)\,d\mu'_\gamma(g) = \sum_\Gamma \iint \varphi(gx)\,d\pi(x)\,d\mu'_\gamma(g) = \\
&\iint \varphi(gx)\,d\pi(x)\,d\mu(g) = \int \varphi(x)\,d\pi'(x)
\end{aligned}$$

This completes the proof.

In the remainder of our discussion c_1, c_2, c_3, etc. will denote various constants. For any given index i, the constant c_i will vary according to the context.

Lemma 6.2. Let π be a Γ-invariant measure on $\mathbb{R}^2 - \{0\}$ and let $f(x) = f_0(\|x\|)$ where f_0 is a monotonic function on $(0,\infty)$. There exist constants c_1, c_2, c_3 such that

$$\int_{r_1<\|x\|<r_2} f(x)\,d\pi(x) < c_3 \int_{c_1 r_1<\|x\|<c_2 r_2} f(x)\,dm(x) = 2\pi c_3 \int_{c_1 r_1}^{c_2 r_2} r f_o(r)\,dr.$$

c_1, c_2, c_3 depend only on π and not on $f(x)$ or r_1, r_2.

Proof: If M is a fixed compact set of G then clearly $c_4\|x\| \leq \|gx\| \leq c_5\|x\|$ for all $x \in \mathbb{R}^2 - \{0\}$ and $g \in M$. Take M to be the support of μ in the foregoing lemma. We shall give the argument for f_o a monotone decreasing function, the idea being similar in the reverse case. Denote by χ_{r_1,r_2} the characteristic function of $[r_1, r_2] \subset (0,\infty)$. We have, for $g \in M$,

$$\int_{r_1<\|x\|<r_2} f(x)\,d\pi(x) = \int \chi_{r_1,r_2}(\|x\|) f_o(\|x\|)\,d\pi(x) \leq$$

$$\int \chi_{c_4 r_1, c_5 r_2}(\|gx\|) f_o(c_5^{-1}\|gx\|)\,d\pi(x).$$

Hence, integrating over $g \in M$

$$\int_{r_1<\|x\|<r_2} f(x)\,d\pi(x) \leq \mu(M)^{-1} \iint \chi_{c_4 r_1, c_5 r_2}(\|gx\|) f_o(c_5^{-1}\|gx\|)\,d\pi(x)\mu(g) =$$

$$\mu(M)^{-1} \int \chi_{c_4 r_1, c_5 r_2}(\|x\|) f_o(c_5^{-1}\|x\|)\,dm(x)$$

by Lemma 6.1. The latter integral may be written

$$c_6 \int_{c_4 r_1}^{c_5 r_2} f_o(c_5^{-1} r)\,r\,dr = c_7 \int_{c_5^{-1} c_4 r_1}^{r_2} f_o(r)\,r\,dr$$

which yields the assertion of the lemma.

Lemma 6.3. Let $g_N = \begin{pmatrix} N & 0 \\ 0 & N^{-1} \end{pmatrix}$ for $N \geq 1$ and let Q denote the region $c_1 \leq \|x\| \leq c_2$ in $\mathbb{R}^2 - \{0\}$ with χ_Q denoting its characteristic function. Set

$$H_N(x) = \int_0^{2\pi} \chi_Q(g_N k_\theta x)\,d\theta.$$

then $H_N(x) = 0$ for $\|x\| < c_1 N^{-1}$ and for $\|x\| > c_2 N$. For all x

we have $\underset{N}{H}(x) \leq C\, N^{-1}\|x\|^{-1}$ for some c.

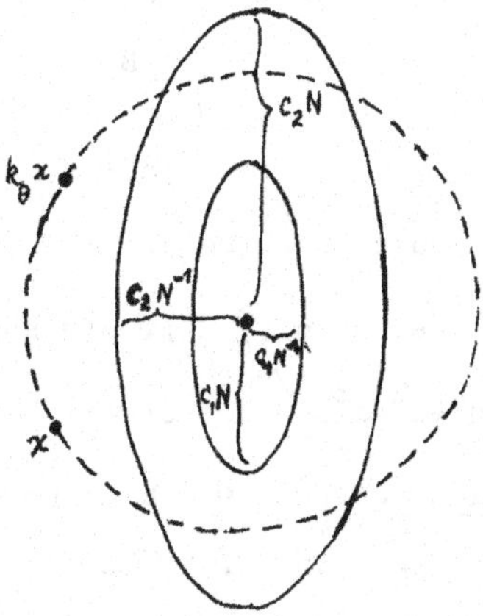

Proof: The proof is apparent from the diagram.

Lemma 6.4. With g_N and Q as in the foregoing lemma, assume φ is a bounded non-negative function with support in Q. Then

(6.1) $$\iiint \varphi(g_N k_\theta x)\, \varphi\, (g_N k_\theta y)\, d\theta\ \frac{d\pi(x)}{\|x\|^2}\, \frac{d\pi(y)}{\|y\|^2}$$

is bounded independently of N.

Proof: To begin with

$$\frac{1}{2\pi}\int_0^{2\pi} \varphi(g_N k_\theta x)\varphi\, (g_N k_\theta y)\, d\theta \leq \{\frac{1}{2\pi}\int_0^{2\pi} \varphi^2(g_N k_\theta x)\, d\theta\}^{1/2}\{\frac{1}{2\pi}\int_0^{2\pi} \varphi^2(g_N k_\theta y) d\theta\}^{1/2}$$

$$\leq c_3\{\int_0^{2\pi} \chi_Q(g_N k_\theta x)\, d\theta\}^{1/2}\{\int_0^{2\pi} \chi_Q(g_N k_\theta y)\, d\theta\}^{1/2}$$

$$= c_3\ \sqrt{H_N(x)\, H_N(y)}\,.$$

this vanishes unless both $\|x\|$ and $\|y\|$ exceed $c_1 N^{-1}$ and in any case it is less than $c_4 N^{-1}\ \sqrt{\|x\|^{-1}\|y\|^{-1}}$. Hence the expression in (6.1) is bounded by

$$c_4 N^{-1} \iint_{c_1 N^{-1} < \|x\|, \|y\| < c_2 N} \|x\|^{-5/2}\, \|y\|^{-5/2}\, d\pi(x)\, d\pi(y)$$

Thus it suffices to show that

$$\int_{c_1 N^{-1} < \|x\| < c_2 N} \|x\|^{-5/2}\, d\pi(x)$$

But by Lemma 6.2, this integral is comparable with

$$\int_{c_5 N^{-1}}^{c_6 N} r^{-5/2},\ r dr = O(\sqrt{N})$$

this proves the lemma.

Let $g = \begin{pmatrix} a & b \\ c & d \end{pmatrix}$. We shall denote by $N(g)$ the number $(a^2+b^2+c^2+d^2)^{1/2}$. Clearly $N(k_\theta g k_{\theta'}) = N(g)$.

<u>Lemma 6.5.</u> Let φ be a non-negative function of compact support on $\mathbb{R}^2 - \{0\}$. For $x, y \in \mathbb{R}^2 - \{0\}$ set

$$H_u(x,y) = \int_{N(g)<u} \varphi(gx)\,\varphi(gy)\,dg$$

and

$$V_u = \int_{N(g)<u} dg.$$

Then

$$\iint H_u(x,y)\,\frac{d\pi(x)}{\|x\|^2}\,\frac{d\pi(y)}{\|y\|^2} = O(V_u) \tag{6.2}$$

as $u \longrightarrow \infty$.

<u>Proof:</u> Let $F(g)$ be any bounded borel function on G. The region $\{N(g) < u\} \subset G$ is invariant under right and left multiplication by K. Since every group element can be expressed in precisely two ways in the form $k_\theta a_s k_{\theta'}$, $k_\theta, k_{\theta'} \in K$, $a_s \in A$, we can express

$$\int_{N(g)<u} F(g)\,dg = c_1 \int_{N(a_s)<u}\int_0^{2\pi}\int_0^{2\pi} F(k_\theta a_s k_{\theta'})\,d\theta\,d\theta'\,d\eta(t)$$

for some measure η on $(0,\infty)$. Now both V_u and $H_u(x,y)$ are integrals of the above type and for both $F(k_\theta g) = F(g)$. Hence (6.2) will follow if it is known that

$$\iiint \varphi(gk_\theta x)\,\varphi(gk_\theta y)\,\frac{d\pi(x)}{\|x\|^2}\,\frac{d\pi(y)}{\|y\|^2} < c_2$$

for all diagonal matrices g. But this is the content of Lemma 6.4.

The next lemma deals with G-invariant area on $\mathfrak{D}$. We denote by $D(z, z')$ the G-invariant metric on $\mathfrak{D}$ and by $\mathfrak{D}_R$ the disc of D-radius R about 0. It can be shown that the area of $\mathfrak{D}_R$ is given by

$$A_R = \sinh^2 R/2 .$$

We use this in

<u>Lemma 6.6.</u> For fixed a there exists c with $A_{R+a} < cA_R$ for all R.

Lemma 6.6 can be used to prove

<u>Lemma 6.7.</u> Let V_u be as in Lemma 6.5 and let a be fixed. There exists c with $V_{au} < c\,V_u$.

<u>Proof</u>: Let W be a subset of G-invariant under right multiplication by K. Because of the uniqueness, up to constant multiple, of the G-invariant measure on $\mathfrak{D}$, it follows that the volume of G in W can be taken to equal to the area of $W(0) \subset \mathfrak{D}$. For $W = \{g \mid N(g) < u\}$ we find $W(0)$ equal to a disc of radius $R(u)$. (It is not necessary to compute this.) Now any matrix with $N(g) < ua$ can be obtained by multiplying a matrix with $N(g) < u$ by one in a fixed compact set, and since

$$\begin{aligned} D(g_1g_2(0),0) &\leq D(g_1g_2(0),\ g_1(0)) + D(g_1(0),0) \\ &= D(g_2(0),0) + D(g_1(0),0) \end{aligned}$$

we find that $R(ua) < R(u) + a'$. So the lemma follows from Lemma 6.6.

<u>Lemma 6.8.</u> With $x, y \in \mathbb{R}^2 - \{0\}$, φ a fixed function of compact support in $\mathbb{R}^2 - \{0\}$, set

$$H_u'(x,y) = \sum_{\substack{\gamma \in \Gamma \\ N(\gamma) < u}} \varphi(\gamma x)\varphi(\gamma y)$$

then

$$\iint H_u'(x,y) \frac{d\pi(x)}{\|x\|^2} \frac{d\pi(y)}{\|y\|^2} = O(V_u)$$

<u>Proof</u>: Let M be a fundamental domain for Γ in G; i.e., a borel set with $M_\gamma \cap M = \emptyset$ for $\gamma \neq e$, $r \in \Gamma$, and $\bigcup_{\Gamma} M\gamma = G$. We may assume that $\overline{M}$ is compact. Set $\Phi(x) = \max\{|\varphi(q^{-1}x)| \mid q \in \overline{M}\}$ so that

Φ is bounded and has compact support. If $g \in q\gamma$, $q \in M$, $\gamma \in M$ and if $N(\gamma) < u$, then $N(g) < cu$ for some constant c. It follows that if $F(g)$ is a non-negative borel measurable function on G, then

$$\sum_{\substack{N(\gamma)<u \\ \gamma\in\Gamma}} \int_M F(q\gamma)\,dq \leq \int_{N(g)<cu} F(g)\,dg.$$

Also $|\varphi(\gamma x)\varphi(\gamma y)| \leq \Phi(q\gamma x)\Phi(q\gamma y)$ for $q \in M$.

Hence

$$m_G(M) \left| \sum_{N(\gamma)<u} \varphi(\gamma x)\varphi(\gamma y) \right| \leq \int_{N(g)<cu} \Phi(gx)\Phi(gy)\,dg$$

and we may apply Lemma 6.5 to obtain the desired result.

Lemma 6.9. Let the harmonic function $h_\varphi(z)$ be defined by

$$h_\varphi(gK) = \int \varphi(x)\,\|g^{-1}x\|^{-2}d\pi(x)$$

where φ is a non-negative function of compact support in $\mathbb{R}^2 - \{0\}$. Then

$$\sum_{\gamma(0)\in \mathcal{D}_R} h_\varphi(\gamma(0))^2 = O(A_R)$$

where A_R is the non-euclidean area of the disc $\mathcal{D}_R$.

Proof: The condition $g(0) \in \mathcal{D}_R$ is equivalent to a condition of the form $N(g) < u$. We use the Γ-invariance of π to write

$$h_\varphi(\gamma(0)) = \int \varphi(\gamma x)\,\|x\|^{-2}d\pi(x)$$

so that

$$\sum_{\gamma(0)\in \mathcal{D}_R} h_\varphi(\gamma(0))^2 = \iint H_u'(x,y)\,\frac{d\pi(x)}{\|x\|^2}\,\frac{d\pi(y)}{\|y\|^2}$$

for appropriate u, with H_u' as in Lemma 6.8. Bearing in mind the connection between areas in $\mathcal{D}$ and volumes in G we see that this lemma follows from Lemma 6.8.

Lemma 6.10. Let M be a compact set in G. There exists a constant $c = c(M)$ such that for any positive harmonic function on G/K

$$h(gg_oK) \leq c\,h(gK)$$

for $g \in G$, $g_o \in M$.

Proof: It suffices to prove this for extremal harmonic functions $P(\zeta, z)$, or using the form in Proposition 3.1, it suffices to consider the functions $\|g^{-1}x\|^{-2}$. But

$$\|g_0^{-1}g^{-1}x\|^{-2} \leq \|g_0\|^2 \|g^{-1}x\|^{-2}$$

and $\|g_0\|$ is bounded in M.

Lemma 6.11. With h_φ as in Lemma 6.9 we have

$$\int_{\mathcal{D}_R} h_\varphi(z)^2 dm(z) = O(A_R)$$

Proof: We let Q be a bounded fundamental domain for Γ in $\mathcal{D}$ with $o \in Q$. $\bigcup_\Gamma \gamma Q$ covers $\mathcal{D}$ and since Q is bounded we find that each $z \in \mathcal{D}_R$ can be expressed as $\gamma z'$, $\gamma \in \Gamma$, $z' \in Q$ and

$$D(\gamma(0), 0) \leq D(\gamma z', \gamma(0)) + D(\gamma z', 0) \leq D(z', 0) + R.$$

Hence $\gamma(0) \in \mathcal{D}_{R+c}$ for appropriate c independent of R. So

$$\int_{\mathcal{D}_R} h_\varphi(z)^2 \, dm(z) \leq \sum_{\substack{\gamma \in \Gamma \\ \gamma(0) \in \mathcal{D}_{R+c}}} \int_Q h_\varphi(\gamma z')^2 dm(z').$$

Now $z' \in Q$ can be expressed $z' = g'(0)$ with g' inside some compact set. Hence by the foregoing lemma

$$h_\varphi(\gamma z') = h_\varphi(\gamma g'(0)) < c_1 \, h_\varphi(\gamma(0)).$$

This gives

$$\int_{\mathcal{D}_R} h_\varphi(z)^2 dm(z) < c_2 \sum_{\substack{\gamma \in \Gamma \\ \gamma(0) \in \mathcal{D}_{R+c}}} h_\varphi(\gamma(0))^2 = O(A_{R+c}) = O(A_R).$$

Finally we argue that the integral in the foregoing lemma is a weighted linear combination of integrals $I_r = \int h_\varphi(re^{i\theta})^2 \, d\theta$ for various r. Since the weight given to I_r for $r_1 < r < r_2$ depends on the non-euclidean area of this annulus, it follows that the total weight given to circles with $r < r_1 < 1$ tends to 0 in the integral over $\mathcal{D}_R$ as $R \longrightarrow \infty$ (since $A_R \longrightarrow \infty$). In view of the fact that the I_r are increasing as $r \nearrow 1$ it follows from Lemma 6.11 that I_r must be bounded. We have proven

Proposition 6.12. If π is a Γ-invariant measure on $\mathbb{R}^2 - \{0\}$ and $\varphi(x)$ is a function of compact support on $\mathbb{R}^2 - \{0\}$ then

$$h_{\varphi}(gK) = \int \varphi(x) \| g^{-1}x \|^{-2} d\pi(x)$$

is an L^2 - bounded harmonic function.

Corollary. The horocycle flow is uniquely ergodic.

BIBLIOGRAPHY

[1] L. Auslander, L. Green, and F. Hahn, Flows on Homogeneous Spaces. Annals of Mathematics Studies No. 53 (1963), Princeton.

[2] H. Furstenberg, "Strict ergodicity and transformations of the torus", American Journal of Mathematics 83 (1961) 573-601

[3] H. Furstenberg and I. Tzkoni, "Spherical functions and integral geometry", Israel Journal of Mathematics, 10 (1971) 327-338.

[4] G. A. Hedlund, "On the metrical transitivity of the geodesics on closed surfaces of constant negative curvature", Annals of Mathematics 35 (1934) 787-808.

[5] G. A. Hedlund, "A metrically transitive group defined by the modular group", Americal Journal of Mathematics, vol. 57 (1935) 668-678.

[6] G. A. Hedlund, "Fuchsian groups and transitive horocycles", Duke Mathematical Journal 2 (1936) 530-542.

[7] E. Hopf, Ergodentheorie, Ergebnisse der Mathematik und ihrer Grenzgebiete 5 (1937).

[8] C. C. Moore, "Ergodicity of flows on homogeneous spaces", American Journal of Mathematics 88 (1965) 154-178.

[9] A. Zygmund, Trigonometric Series , vol. I second edition, 1959, Cambridge.

Berkeley, California
and
Jerusalem, Israel

How Should We Define Topological Entropy?

L. Wayne Goodwyn

Let X be a compact topological space and let ϕ be a continuous map from X into itself. In [1], the following definitions were made.

For an open cover $\mathcal{U}$ of X, let $N(\mathcal{U})$ be the minimum number of members of $\mathcal{U}$ needed to cover X. Let $H(\mathcal{U}) = \log N(\mathcal{U})$. Let

$$h(\phi, \mathcal{U}) = \lim_{n\to\infty} \frac{1}{n} H\left(\bigvee_{i=0}^{n-1} \phi^{-i}\mathcal{U}\right),$$

and finally, let

$$h(\phi) = \sup\{h(\phi, \mathcal{U}) \mid \mathcal{U} \text{ is an open cover of } X\}.$$

The point of view taken is that X is a state space and that open covers represent experiments, the sets of a cover being its outcomes. The map ϕ represents discrete time. The quantity $H(\mathcal{U})$ is intended to represent the amount of information obtained by performing the experiment $\mathcal{U}$. The quantity $h(\phi, \mathcal{U})$ is then logically interpreted as the long time average of the amount of information obtained per performance of $\mathcal{U}$, if it is performed many times in succession. The topological entropy, $h(\phi)$, is then interpreted as the largest possible rate at which information can continually be extracted from the system, and hence a measure of the unpredictability, or randomness of the system.

If topological entropy is to be viewed as a good definition of randomness, rather than merely an imitation of "measure-theoretic" or "probabalistic" entropy (See [2]), then the above interpretations must be justified.

There are several places where disagreement could occur as to the validity of these interpretations. One such would be the interpretation of open covers as experiments. It would seem natural to view a partition as an experiment. $N(\mathcal{U})$ can be defined for finite partitions, and thus the entropy could be defined as the supremum of $h(\phi, \mathcal{U})$ over all finite partitions. There is, in fact, considerable justification for the use of open covers, but it will not be presented here.

We propose here to shed a little light on the more obviously doubtful viewpoint: that $H(\mathcal{U})$ represents information. In other words, perhaps a different definition of $N(\mathcal{U})$, and hence $H(\mathcal{U})$ should have been used. In fact, there is a perfectly reasonable alternative. It was introduced by R. Bowen, [3], in a different form and setting.

Definition: Let $N^*(\mathcal{U})$ be the maximal number of points in a subset F of X with the property that no two distinct points of F lie in a single set of $\mathcal{U}$.

It is then possible to define a different kind of entropy, $h^*(\phi)$, using N^* in the place of N. (However, the limit in the definition of $h(\phi, \mathcal{U})$ must be replaced by limit-infimum.)

Although there are examples which show that $N^*(\mathcal{U})$ may be different from $N(\mathcal{U})$, and, in fact, that $h^*(\phi, \mathcal{U})$ may be different from $h(\phi, \mathcal{U})$, it so happens that $h^*(\phi) = h(\phi)$ if X is assumed to be Hausdorff. (The proof will appear elsewhere.)

We now turn to the problem of other possible definitions for the notion of the information of a cover. At this point, we accept the usual definition as valid if the cover is assumed to be essential. An essential cover is one in which the deletion of any one member will result in a collection that does not cover. For essential covers, it seems rather unlikely that a better definition for information than $\log N(\mathcal{U})$ can be found. In this case $N(\mathcal{U})$ is the same as $N^*(\mathcal{U})$, which is just the number of members in the cover. (Perhaps the validity does need justification even for essential covers, but the problem will not be pursued here.)

The chief concern seems to be, then, that of defining "information" for nonessential covers, and, of course, $\bigvee_{i=0}^{n-1} \phi^{-i}\mathcal{U}$ may be non-essential even when $\mathcal{U}$ is essential, so that the problem is not easily avoided. The following theorem provides an answer. (The proof will appear elsewhere.)

Theorem: Let M be a function which assigns to each open cover $\mathcal{U}$ of X a real number $M(\mathcal{U})$ such that

1. If $\mathcal{U}$ is a refinement of $\mathcal{V}$, then $M(\mathcal{U}) \geq M(\mathcal{V})$.
2. If $\mathcal{U}$ is essential, then $M(\mathcal{U}) = N(\mathcal{U})$.

Then $N^*(\mathcal{U}) \leq M(\mathcal{U}) \leq N(\mathcal{U})$ for all open covers $\mathcal{U}$ of X.

If we were to use M to define another kind of entropy, $h_M(\phi)$, by using M in the place of N, this theorem combined with the fact that $h^*(\phi) = h(\phi)$ yields $h_M(\phi) = h(\phi)$.

Thus, in a sense, it does not matter how we define "information" of a cover; it will all come out in the wash.

References

1. R. Adler, A. Konheim and M. McAndrew, Topological Entropy, Trans. Amer. Math. Soc. 114(1965), 309-319.

2. P. Billingsley, Ergodic Theory and Information, Wiley New York, 1965.

3. R. Bowen, Entropy for group endomorphisms and homogeneous spaces, Trans. Amer. Math. Soc. 152(1971), 401-414.

University of Kentucky
Lexington, Kentucky

SOME GENERAL DYNAMICAL NOTIONS

Walter Gottschalk
Wesleyan University

The following are familiar standard notions for flows: (I) recurrent points under flows; (II) symbolic flows; (III) embedding a discrete flow in a continuous flow by enlarging the phase space. I wish to comment on ways of generalizing these notions from flows to transformation groups.

(I) Consider the standard notion of recurrent point under a flow. How can we systematically explore the possibilities of generalizing this notion from flows to transformation groups? Here is one approach.

A cluster on a set X is defined to be a class of subsets of X. A cluster $\mathcal{A}$ on a uniform space $(X, \mathcal{U})$ is said to be $\mathcal{U}$-round or round provided that $\forall A \in \mathcal{A}. \exists B \in \mathcal{A}. \exists \alpha \in \mathcal{U}. A \supset B\alpha$. If $\mathcal{A}$ is a cluster on a uniform space $(X, \mathcal{U})$, then the round hull of $\mathcal{A}$ is defined to be the round cluster $\{A\alpha \mid A \in \mathcal{A} \ \& \ \alpha \in \mathcal{U}\}$ on X.

Let (X, T) be a transformation group, let $\mathcal{A}$ be a cluster on T, let $\mathcal{B}$ be the left round hull of $\mathcal{A}$, and let $x \in X$. Then x is $\mathcal{A}$-recursive iff x is $\mathcal{B}$-recursive.

Let X be a set, and let $\mathcal{F}$ be a cluster on X. A subset A of X is said to be $\mathcal{F}$-intersective provided that $\forall F \in \mathcal{F}. A \cap F \neq \emptyset$. The class of all $\mathcal{F}$-intersective subsets of X is denoted $\mathcal{F}^{\delta}(X)$ or $\mathcal{F}^{\delta}$.

Let (X, T) be a transformation group, let $\mathcal{F}$ be a cluster on T, let $\mathcal{G}$ be the left round hull of $\mathcal{F}$, and let $x \in X$. Then x is $\mathcal{F}^{\delta}$-recursive iff x is $\mathcal{G}^{\delta}$-recursive.

Let T be a Hausdorff topological group with identity element e, and let (Se, T) be the greatest ambit under T. It is possible [1] to think of (Se, T) as given by a compactification of T in the following way: let S be the completion of T with respect to the left Samuel uniformity of T, consider $e \in T \subset S$, and let T act upon S by continuous extension of the right translations of T.

Some of the research represented in this paper was performed while the author was partially supported by National Science Foundation Grant GP-18825.

Let (X, T) be a transformation group with compact Hausdorff phase space X, and let $x \in X$. Since the greatest ambit Se is uniquely morphic to the ambit $\overline{xTx}$, whatever is done in S is automatically reflected in $\overline{xT}$.

Let L be a nonempty closed invariant subset of S and let $\mathcal{F}$ be (the left round hull of) the trace on T of the neighborhood filter of L in S. Then:

(1) $\mathcal{F}$ is an invariant (left round) filter on T.

(2) The adherence of $\mathcal{F}$ in S is L.

(3) If Xx is an ambit, and if $\varphi : Se \to Xx$ is the unique morphism, then: (i) $L\varphi$ = the adherence of $x\mathcal{F}$ in X and is denoted L_x; (ii) x is $\mathcal{F}^\delta$-recursive iff $x \in L_x$.

Conversely, let $\mathcal{F}$ be an invariant filter on T, and let L be the adherence of $\mathcal{F}$ in S. Then L is a nonempty closed invariant subset of S and maps onto the adherence L_x of the filter-base $x\mathcal{F}$ on X.

Thus whether starting with $\mathcal{F}$ or L, the other can be obtained, although it is not clear at the moment what is needed to make the correspondence one-to-one.

Let $\mathcal{F}$ be an invariant filter on T. Then x is defined to be $\mathcal{F}$-recurrent provided that x is $\mathcal{F}^\delta$-recursive.

Let $F = (\mathcal{F}_i \mid i \in I)$ be a family of invariant filters on T. A subset A of T is said to be F-<u>intersective</u> in case A is $\mathcal{F}_i$-intersective for every $i \in I$. Let F^δ denote the class of all F-intersective subsets of T. Then x is defined to be F-recurrent provided that x is F^δ-recursive. Note that x is F-recurrent iff x is $\mathcal{F}_i$-recurrent for all $i \in I$.

If $\mathcal{F}$ is an invariant filter on T, then each member of $\mathcal{F}$ is discretely left replete. If $\mathcal{F}$ is an invariant left round filter on T, then each member of $\mathcal{F}$ is left replete. If $P \subset T$, then $\{Pt \mid t \in T\}$ is a filter subbase on T iff P is discretely left replete.

The notion of recurrence in [2] is based upon the family of filters of translates of replete semigroups and is thus a special instance of the above.

Evidently, the customary notions of positively recurrent point, negatively recurrent point, unilaterally recurrent point, and bilaterally recurrent point under a flow are subsumed under the general notion of recurrent point as given above. Here is another application for flows. For definiteness, consider a continuous flow and positive recurrence. Let $\mathfrak{F}$ be the filter on $\mathbb{R}$ of all subsets of $\mathbb{R}$ whose ω-density (that is, in $\mathbb{R}[0, +\infty])$ is 1. Then $\mathfrak{F}^{\delta}$ consists of those subsets of $\mathbb{R}$ whose upper ω-density is positive. The corresponding notion of recurrence may be called _strong_ recurrence.

(II) It is possible to generalize at least some of symbolic dynamics to certain kinds of transformation groups.

Let P be a finite set with at least two elements, let P be provided with its discrete topology, let T be an infinite discrete group, let X be the cartesian power P^T provided with its product topology, and let T act upon X by left translation. Then (X, T) is called the _left symbolic transformation group over_ T _to_ P. If T is the additive group $\mathbb{Z}$ of integers, then (X, T) is the standard symbolic flow. In general, X is compact Hausdorff zero-dimensional self-dense, and (X, T) is expansive. A presumably large project is to correlate group properties of T with dynamical properties of (X, T). Here are some recent results of Wayne Lawton [5, 6] in this context:

(1) T is profinite iff the set of periodic points of (X, T) is dense in X.

(2) Call T _surjunctive_ in case every one-to-one endomorphism of (X, T) is onto for all P. If T is locally finite or profinite or abelian, then T is surjunctive. Also every subgroup of a surjunctive group is surjunctive.

No example of a non-surjunctive group seems to be known. If it could be proved that every quotient group of a surjunctive group is surjunctive, then every group would be surjunctive.

Professor Hedlund has pointed out that every symbolic flow is surjunctive [3].

(III) The following is complementary to the paragraphs 1.43, 1.44, 1.45 of [2].

If (X, T) is a transformation group, then X/T denotes the orbit partition $\{xT \mid x \in X\}$ of X under T provided with its partition topology. If $(X, T \times S)$ is a transformation group, then $(X, T \times e_S)$ and $(X, e_T \times S)$ are commuting transformation groups and the transformation group $(X/(T \times e), e \times S)$ is defined [2, 1.40].

Let (X, T) be a transformation group where X is an arbitrary topological space and T is an arbitrary topological group, and let S be a topological group such that T is a topological subgroup of S. The question is to isomorphically embed the transformation group (X, T) in some transformation group (Y, S). Define the transformation group $(X \times S) \times (T \times S) \to X \times S$, $(x, \sigma)(t, s) = (xt, t^{-1}\sigma s)$. The transformation group $((X \times S)/(T \times e), e \times S)$ is called the S-<u>extension</u> <u>of</u> (X, T). Consider the canonical injection $X \to X \times S$, $x \mapsto (x, e)$, consider the canonical projection $X \times S \to (X \times S)/(T \times e)$, $(x, \sigma) \mapsto \{(x\tau, \tau^{-1}\sigma) \mid \tau \in T\}$, and consider their composite $\varphi : X \to (X \times S)/(T \times e) = Y$, $x \mapsto \{(x\tau, \tau^{-1}) \mid \tau \in T\}$. Then:

(1) $\varphi : (X, T) \to (Y, T)$ is one-to-one continuous T-equivariant.

(2) $X\varphi T = X\varphi$, $X\varphi S = Y$.

(3) $X\varphi \cap X\varphi(S \dashv T) = \emptyset$.

(4) $X\varphi$ is closed in Y iff T is closed in S.

(5) If X is Hausdorff and if T is closed in S, then Y is Hausdorff.

(6) If X is compact and if T is syndetic in S, then Y is compact.

(7) If T is discrete or if X is compact Hausdorff and T is closed in S, then $\varphi : (X, T) \to (Y, S)$ is an isomorphic embedding.

In the language of fiber bundles [4], $(X \times S)/(T \times e) \to S/T$, $\{(x\tau, \tau^{-1}\sigma) \mid \tau \in T\} \mapsto T\sigma$, is a fiber bundle over S/T with fiber X, structure group T, and associated principal bundle $S \to S/T$.

Let (Y, S) be a transformation group where Y is an arbitrary topological space and S is an arbitrary topological group, let T be a syndetic subgroup of S, and let X be a closed subset of Y such that $XT = X$, $XS = Y$ and $X \cap X(S \dashv T) = \emptyset$. Then (Y, S) is canonically isomorphic to the S-extension of (X, T), the canonical isomorphism being established by the map $X \times S \to Y$, $(x, s) \mapsto xs$.

The transformation group $(X \times S) \times (T \times S) \to X \times S$, $(x, \sigma)(t, s) = (xt, t^{-1}\sigma s)$, as defined above can be analyzed in various ways by using different kinds of product transformation groups which are described in generality below.

If $X_i \times T_i \to X_i$, $i \in I$, is a family of transformation groups, if T is a topological group, and if $\varphi_i : T \to T_i$ is a continuous group homomorphism for each $i \in I$, then

$$(\times_{i \in I} X_i) \times T \to \times_{i \in I} X_i$$

$$(x_i \mid i \in I)t = (x_i \cdot t\varphi_i \mid i \in I)$$

is a transformation group. This definition can be specialized in three ways: (i) if $X \times T_0 \to X$ is a transformation group, if T_0 is a topological group, and if $\varphi : T \to T_0$ is a continuous group homomorphism, then $X \times T \to X$, $xt = x \cdot t_0\varphi$, is a transformation group; (ii) if $X_i \times T_i \to X_i$, $i \in I$, is a family of transformation groups, then $pr_j : \times_{i \in I} T_i \to T_j$ is a continuous group homomorphism for each $j \in I$ and

$$(\times_{i \in I} X_i) \times (\times_{i \in I} T_i) \to \times_{i \in I} X_i$$

$$(x_i \mid i \in I)(t_i \mid i \in I) = (x_i t_i \mid i \in I)$$

is a transformation group; (iii) if $X_i \times T \to X_i$, $i \in I$, is a family of transformation groups, then

$$(\times_{i \in I} X_i) \times T \to \times_{i \in I} X_i$$

$$(x_i \mid i \in I)t = (x_i t \mid i \in I)$$

is a transformation group.

If $X \times T \to X$ and $X \times S \to X$ are transformation groups, then $X \times (T \times S) \to (X \times T) \times S \to X \times S \to X$ is a transformation group iff the actions of T and S commute, that is, $\forall x \in X. \forall t \in T. \forall s \in S. xts = xst$. If n is a positive integer, if $X \times T_i \to X$, $i \in \{1, \cdots, n\}$, is a family of transformation groups, then $X \times (\times_{i \in I} T_i) \to X$, $x(t_1, \cdots, t_n) = xt_1 \cdots t_n$, is a transformation group iff the actions of $T_1, \cdots, T_n$ pairwise commute.

REFERENCES

1. Brook, Robert B., A construction of the greatest ambit, Mathematical Systems Theory, vol. 4 (1970), pp. 243-248.

2. Gottschalk, Walter H.; Hedlund, G. A., Topological dynamics, American Mathematical Society Colloquium Publications, vol. 36. American Mathematical Society, Providence, 1955. Reprinted with corrections and an appendix, 1968.

3. Hedlund, G. A., Endomorphisms and automorphisms of the shift dynamical system, Mathematical Systems Theory, vol. 3 (1969), pp. 320-375.

4. Husemoller, Dale, Fibre bundles, McGraw-Hill, New York, 1966.

5. Lawton, Wayne M., Expansive transformation groups, Ph.D. dissertation, Wesleyan University, 1972.

6. Lawton, Wayne M., Note on symbolic transformation groups, Notices of the American Mathematical Society, vol. 19 (1972), p. A-376 (abstract).

GROUP-LIKE DECOMPOSITIONS OF RIEMANNIAN BUNDLES

Leon W. Green
University of Minnesota*)

§0. INTRODUCTION

Horocycles, at least in the hyperbolic plane, have been studied for many years. However, it was Hedlund who first used them in topological dynamics and ergodic theory ([9], [10]), and it was he who had the courage actually to apply them in the case of non-constant curvature. It is this assurance, subsequently exploited by E. Hopf, Anosov, and Sinaĭ ([12, [13], [1], [2]), that has led me, through the back door, so to speak, to a viewpoint of Riemannian geometry about which I intend to speak today. The claim is that arbitrary Riemannian manifolds have much in common with Riemannian symmetric spaces, and that horocycles enable us to construct analogues of the Iwasawa and Bruhat decompositions of semisimple Lie groups when the curvature is strictly negative.

In §1 we establish the notation and express the Cartan structure equations in vector form. The relationship to the Cartan decomposition of a semisimple group is pointed out, and meanings attached to the assertion that the resulting flows act (pointwise) transitively in the bundle of frames. In §2 we recall the definition of horospheres and give growth estimates for certain functions associated with them. §3 is devoted to defining the horocycles. §4 is qualitative, pointing out the analogies to the Iwasawa and Bruhat decompositions connected with rank one symmetric spaces. In §5 we give an application to ergodic theory of these ideas, namely, the

Theorem. Let M be a compact Riemannian manifold with 1/4-pinched negative curvature. Then any generalized geodesic flow in the bundle of orthonormal frames is ergodic.

---

*) Part of this work was done while the author was visiting the University of California at San Diego. Partial support came from NSF GP-27670.

It then follows from a theorem of Ken Thomas that this flow is a Kolmogorov flow, although it is clear that when the dimension is greater than two, it is not an Anosov flow.

Complete proofs will appear elsewhere.

§1. STRUCTURE EQUATIONS AND THE CARTAN DECOMPOSITION

Let $\mathbb{F}M$ (abbreviated: $\mathbb{F}$) be the bundle of orthonormal frames of the complete, n-dimensional, connected, C^∞ Riemannian manifold M. $\mathbb{F}$ is a principal $O(n, \mathbb{R})$-bundle over M, and each $f \in \mathbb{F}$ may be interpreted as an isomorphism of $\mathbb{R}^n$ onto $T_{\pi(f)}M$, the tangent space of M at the point $\pi(f)$. (π is the projection of $\mathbb{F}$ onto M.) For $f \in \mathbb{F}$, $\xi \in \mathbb{R}^n$, and $g \in O(n, \mathbb{R})$, the right-action, R_g, of the group on $\mathbb{F}$ obeys the rule $(R_g f)\xi \equiv (fg)\xi = f(g\xi)$. For brevity, denote $O(n, \mathbb{R})$ by G and its Lie algebra by $\mathfrak{G}$. For $A \in \mathfrak{G}$, the one-parameter group of diffeomorphisms of $\mathbb{F}$, $(f, t) \mapsto f \exp tA$, induces the vector field $\lambda(A)$, called a <u>fundamental</u> vertical vector field. Each $\xi \in \mathbb{R}^n$ of unit length defines $\mathbb{F}$ as an $O(n - 1, \mathbb{R})$ bundle over TM with projection $f \mapsto f\xi$. If $\{\varphi_t\}$ is the geodesic flow in TM, there is a unique horizontal flow in $\mathbb{F}$, $\{\Phi_t^\xi\}$, such that

$$\Phi_t^\xi(f)\xi = \varphi_t(f\xi).$$

The geometric meaning of this is: $\Phi_t^\xi(f)$ is the frame obtained by parallel transporting f along the unit speed geodesic γ a distance t, where γ has the initial condition $\gamma'(0) = f\xi$. Hence for each $\xi \in S^{n-1}$ we obtain a tangent field $B(\xi)$ to the flow $\{\Phi_t^\xi\}$; these and their constant multiples, are called the <u>basic</u> (or <u>standard</u>) horizontal vector fields. The completeness of M assures us that the flows, which we will also call "basic", or <u>generalized geodesic flows</u>, are globally defined one-parameter groups of diffeomorphisms of $\mathbb{F}$. The Cartan structure equations then read

(1.1) $$[\lambda(A),\ B(\xi)] = B(A\xi);$$

(1.2) $$[B(\xi),\ B(\eta)] = \lambda\left(f^{-1}\ R(f\xi,\ f\eta)f\right).$$

Here R is the curvature transformation. If we take the fundamental and basic fields corresponding to bases in $\mathfrak{G}$ and $\mathbb{R}^n$, respectively, we obtain a parallelization of F. The structure equations, together with the fact that λ is an isomorphism, enable us to compute the bracket of any two suitably presented members of $\mathfrak{X}(\mathbb{F})$, the set of all vector fields on $\mathbb{F}$. In general, of course, the structure "constants" of this Lie algebra are not constant.

To use suggestive notation, let $\mathfrak{k}$ denote the image of λ, and $\mathfrak{p}$ the real subspace of $\mathfrak{X}(\mathbb{F})$ spanned by the basic vector fields. Then the structure equations say that $[\mathfrak{k},\ \mathfrak{p}] \subseteq \mathfrak{p}$ and $[\mathfrak{p},\ \mathfrak{p}]$, even if it is not contained precisely in $\mathfrak{k}$, at least consists only of vertical vector fields. Thus $\mathfrak{X}(\mathbb{F})$ contains a subset of vector fields which carries a structure very much like a reductive Lie algebra, and this subset is large enough to span $\mathfrak{X}(\mathbb{F})$ as a module over $\mathfrak{F}(\mathbb{F})$, the smooth functions on $\mathbb{F}$. The first group-like decomposition of the bundle may then be written

(1.3) $$\mathfrak{X} \approx \mathfrak{k} + \mathfrak{p}.$$

But the analogy with the Cartan decomposition goes further: the flows generated by the basic and fundamental fields actually act transitively on $\mathbb{F}$. This follows directly from the completeness of M. Namely, by considering a geodesic joining $\pi(f)$ and $\pi(f')$, one can find, for arbitrary $f, f' \in \mathbb{F}$, elements $\xi \in S^{n-1}$, $t \in \mathbb{R}$, and $g \in G$ such that $\Phi_t^{\xi} \circ R_g(f) = f'$.

We note for further application another aspect of this transitivity: any locally integrable function which is invariant under the basic and fundamental flows is almost everywhere constant. (The measure on $\mathbb{F}$ is locally the product of the Riemannian volume on M and Haar measure on G.)

§2. HOROSPHERES

From this point on we assume that the sectional curvature of M is strictly negative; i.e., there are constants k and K, with $0 < k < K$, such that, for all unit, mutually orthogonal vectors $u, v \in TM$, $-K^2 \leq \langle R(u, v)v, u\rangle \leq -k^2$. Then the universal covering space of M, $\widetilde{M}$, is diffeomorphic to $\mathbb{R}^n$, is a complete Riemannian manifold, and has the property that any two points may be joined by a unique geodesic segment, the length of which equals the distance between the points. Moreover, any geodesic sphere (the locus of points equidistant from a fixed point) bounds a geodesically strictly convex domain.

Choose an orthonormal basis, $\xi_1, \ldots, \xi_n$, of $\mathbb{R}^n$ and let us abbreviate $B(\xi_n)$ and $\Phi_t^{\xi_n}$ by B and Φ_t, respectively. For any $f \in \mathbb{F}$, set $f(s) = \Phi_s(f)$, $f_n = f\xi_n$, and $f_n(s) = f(s)\xi_n$. Without danger of confusion, we may denote the projection of TM onto M by the same symbol, π, which we use for the projection of $\mathbb{F}$. For $m \in \widetilde{M}$, $r \in \mathbb{R}^+$, let $S(r, m)$ denote the sphere of radius r with center m. For $u \in T\widetilde{M}$, the <u>horosphere</u> with inner normal u is defined to be the limit of the sets $S\big(s, \pi\varphi_s(u)\big)$, as $s \to \infty$. In the case of constant curvature, or even in more general symmetric spaces, these sets have been considered by many authors. This geometric definition has been given by Busemann, Grant, Hedlund, and others.

We proceed to describe some more detailed structure of the horospheres which is needed in the next section. Let $\xi \in \mathbb{R}^n$, $f \in \mathbb{F}$, and $b > 0$. Connect the points $\pi f_n(b)$ and $\pi\varphi_t(f\xi)$ by the geodesic $x_b(s; t, f, \xi)$, where the arc-length parameter s is chosen so that $x_b(b; t, f, \xi) = \pi f_n(b)$, and the segment is traversed as s increases to b. $\frac{\partial}{\partial t} x_b(s; 0, f, \xi)$ is then a Jacobi field along the geodesic $\pi f_n(s)$, so it equals $Z_b(s, f)\xi$, where $Z_b(s, f)$ is the linear map of $\mathbb{R}^n$ into $T_{\pi(f(s))}\widetilde{M}$ satisfying the operator Jacobi equation

$$\nabla_s^2 Z_b(s, f) + R\big(Z_b(s, f)\cdot, f_n(s)\big)f_n(s) = 0.$$

(Here the second term is to be interpreted as the map from $\mathbb{R}^n$ to $T_{\pi(f(s))}\tilde{M}$, evaluated by supplying an argument for the map Z_b.) The boundary conditions on (2.1) are $Z_b(b, f) = 0$, and $Z_b(0, f) = f$. The operator ∇_s is covariant differentiation along the geodesic $\pi f_n(s)$. It is useful also to define the solution $A(s, f)$ of (2.1) which satisfies the boundary conditions $A(0, f) = 0$, $(\nabla_s A)(0, f) = f$. We summarize the properties of these functions in the following

Proposition 2.1. i) $\lim_{c\to\infty} Z_c(s, f) = D(s, f)$ exists, uniformly on compact subsets of $\mathbb{R} \times \mathbb{F}$.

ii) $D(s, f)$ is a solution of (2.1) for which $D(0, f) = f$, and $D(s, f)$ is an isomorphism of vector spaces for all s.

iii) There exists a positive constant C_1 such that $\|A(s, f)x\| \le C_1 \exp(Ks)$ for $s \ge 1$, $\|x\| = 1$.

iv) For any positive $k' < k$, there exist constants C_2, C_3, and s_0 such that, for $s > s_0$,

$$\|x\| \, C_2 \exp(k's) \le \|A(s)x\|,$$

$$\|Z_c(s, f)x\| \le \|x\| \, C_3 \exp(-k's),$$

and consequently $\|D(x, f)x\| \le \|x\| \, C_3 \exp(-k's)$.

Define $U = (\nabla_s D)D^{-1}$. $U(s, f)$ is a non-singular linear transformation on the orthogonal complement of $f_n(s)$ in $T_{\pi(f(s))}\tilde{M}$, and, in fact, is the second fundamental form of the corresponding, horosphere when restricted to that subspace.

Proposition 2.2. i) $U(s, f) = U\big(0, f(s)\big)$.

ii) The spectrum of $U(0, f)$ restricted to the orthogonal complement of f_n lies in $[-k, -K]$.

iii) $U(0, f)$ is continuous.

This proposition follows rather directly by techniques found, e.g., in [8] and Anosov-Sinaǐ [2]. (By a more delicate analysis, B. O'Neill has recently proved iii), even if $k = 0$.) Using the estimates of Proposition 2.1 and a standard variational method, we can improve iii) under additional hypotheses.

Proposition 2.3. Suppose ∇R is bounded and $K^2 < 4k^2$. Then $U(0, f)$ is continuously differentiable.

§3. HOROCYCLES

Hedlund, E. Hopf, Anosov, among others, have amply demonstrated the utility of horospheres in establishing ergodic properties of geodesic flows. What appear to be entirely different proofs have been given, in the symmetric space case, by Gelfand and Fomin [6], and Mautner [14]. In order to show that there is an underlying connection between the two approaches, I wish to define horocycles. Three properties of the horocycles of the hyperbolic plane suggest themselves as candidates for generalization to higher dimensions and not necessarily constant curvature: (a) they have constant geodesic curvature; (b) they are the orthogonal trajectories of asymptotic geodesics; (c) in identifying the unit tangent bundle of the hyperbolic plane with the Mobius group, the vector field which generates the horocycles is an eigenvector field of the adjoint transformation corresponding to the geodesic flow. Grant [7] pointed out, and experience has born her out, that (b) takes precedence over (a) for generalization to surfaces of non-constant curvature. (However, in this regard, see the comments of Anosov, Sinaĭ in [2] and the paper of Arnold, [3].) Our definition of the horospheres in the last section is the natural generalization of (b), and we will use this same method to construct horocycles. However, the chief result is that, by using this definition of horocycles, we automatically obtain vector fields which generalize property (c).

Return to the construction in §2. Let $q_b(s;\, t,\, f,\, \xi)$ be the frame at $x_b(s;\, t,\, f,\, \xi)$ which, when parallel-transported along this latter geodesic to $x_b(b;\, t,\, f,\, \xi)$, coincides with $f(s)$. Hence q_b is obtained from f by parallel translation along the broken path made up of the base geodesic and x_b. Define $Q_b(s;\, f,\, \xi) = \frac{\partial}{\partial t} q_b(s;\, 0,\, f,\, \xi)$. Then

$$\nabla_s Q_b = \nabla_s \nabla_t Q_b = \nabla_t \nabla_s q_b + R\left(f_n(s),\, Z_b(s,\, f)\xi\right) q_b .$$

Since $\nabla_s q_b = 0$, and $q_b(s;\ 0,\ f,\ \xi) = f(s)$, we obtain the differential equation for Q_b

(3.1)
$$\begin{cases} \nabla_s Q_b(s;\ f,\ \xi) = R\left(f_n(s),\ Z_b(s,\ f)\xi\right)\cdot f(s) \\ Q_b(b;\ f,\ \xi) = 0. \end{cases}$$

The unique solution of this system is

$$Q_b(s,\ f,\ \xi) = -\int_s^b R\left(f_n(\sigma),\ Z_b(\sigma,\ f)\xi\right) f(\sigma)\,d\sigma.$$

The estimate (i) of Proposition 2.1 allows us to let b go to infinity, and we define

$$Q(s;\ f,\ \xi) = \lim Q_b(s;\ f,\ \xi) = -\int_s^\infty R\left(f_n(\sigma),\ D(\sigma,\ f)\xi\right) f(\sigma)\,d\sigma.$$

Set $W(s;\ f,\ \eta) = Q\left(s;\ f,\ D^{-1}(s,\ f)f(s)\eta\right)$. Then W is a map from $\mathbb{R}^n$ to $T_{\pi(f(s))}\widetilde{M}$ which satisfies the differential equation

(3.2)
$$(\nabla_s W)(s;\ f,\ \eta) + W\left(s;\ f,\ L^+(s,\ f)\eta\right) + R\left(f(s)\eta,\ f_n(s)\right)\cdot f(s) = 0,$$

where we have set $L^+(s,\ f) = f^{-1}(s)U(s,\ f)f(s)$. This equation is a generalization of the Riccati equation which the function U is known to satisfy. In fact,

(3.3)
$$W(s;\ f,\ \eta)\xi_n = -\,U(s,\ f)f(s)\eta,$$

and

(3.4)
$$W\left(0;\ f(s),\ \eta\right) = W(s;\ f,\ \eta).$$

Theorem 3.1. For each $\eta \in \mathbb{R}^n$ which is orthogonal to ξ_n, define the vector field $H^+(\eta)$ on $\mathbb{F}$ by the equation

$$H^+(\eta) = B(\eta) - \lambda\left(f^{-1}W(0;\ f,\ \eta)\right).$$

Then the fields $\{H^+(\eta)\}$ are continuous and complete, and satisfy

$$[B,\ H^+(\eta)] = -\ H^+(L^+\eta). \tag{3.5}$$

Perhaps the analogy with Lie algebras becomes clearer when we use the basis $\{\xi_i\}$ to set $H_i^+ = H^+(\xi_i)$, $L^+\xi_i = \sum_i L_{ij}^+\xi_j$, $1 \le i,\ j \le n - 1$. Then (3.5) reads

$$[B,\ H_i^+] = -\sum L_{ij}^+(f)H_j^+; \tag{3.6}$$

that is, the submodule spanned by the positive horocycle fields (which is that we call the $H^+(\eta)$'s) is normalized by B, the generalized geodesic flow field.

By applying to $\mathbb{F}$ the involution θ which is induced by the reflection of $\mathbb{R}^n$ in the hyperplane orthogonal to ξ_n, we obtain the negatively directed horocycles, $\{H^-(\eta)\,|\,\eta \perp \xi_n\}$, and equations similar to (3.5) with $L^-(f) = -\ L^+(\theta f)$. Finally, an extension of Proposition 2.3 to the functions W yields

Proposition 3.2. If ∇R is bounded and M is 1/4-pinched, then the horocycle fields are C^1. Hence their associated flows are C^2.

§4. IWASAWA AND BRUHAT DECOMPOSITIONS

It is now evident that Theorem 3.1 yields the decomposition

$$\mathfrak{X} \approx \mathfrak{k} + \mathfrak{a} + \mathfrak{n}, \tag{4.1}$$

where $\mathfrak{a}$ is, of course, the one-dimensional space determined by B and $\mathfrak{n}$ is the subspace spanned by the positive horocycle fields. The corresponding group-like decomposition of $\mathbb{F}$ is the direct analogue of the Iwasawa decomposition for the rank one case. And indeed, $\mathfrak{a}$ is a maximal abelian subspace of $\mathfrak{p}$, since the curvature is bounded away from zero. Then, given $f, f' \in \mathbb{F}$, there exist $g \in G$, $\eta \in \mathbb{R}^{n-1}$, and $s, t \in \mathbb{R}$ such that $R_g \circ \Phi_s \circ X_t f = f'$. To find these elements, one joins $\pi(f')$ by a unique geodesic with the "point of infinity" of the horosphere determined by $f\xi_n$. The intersection of this geodesic with the horosphere determines the point to which f must first be moved by an appropriate horocycle flow, X_t. (The vector field generating $\{X_t\}$ is taken from the submodule spanned by $\mathfrak{n}$.) The parameter s is the value of the Busemann function for $\pi(f')$ relative to the horosphere determined by $f\xi_n$. (See figure 1.)

Let $\mathfrak{m}$ denote the image under λ of $\mathfrak{G}_n$, the subalgebra of $\mathfrak{G}$ which is zero on ξ_n. (G_n, its group, is a copy of $O(n-1, \mathbb{R})$.) Then we have the analogue of the Bruhat decomposition, at least in the rank one special case:

$$\mathfrak{X} \approx \mathfrak{n}^- + \mathfrak{m} + \mathfrak{a} + \mathfrak{n}. \tag{4.2}$$

To obtain the diffeomorphisms commensurate with this decomposition which link two frames f, f', we first construct their associated horospheres, then connect the points at infinity with a geodesic. (This is always possible in the presence of strictly negative curvature--cf., for instance, P. Eberlein and B. O'Neill [5].) Then there exist positive and negative horocycle flows, X_t, X_n^-, a real number s, and an element $m \in G_n$, such that

$$f' = X_r^- R_m' \Phi_s X_t f. \tag{4.3}$$

Here the rotation R_m' may have to be supplemented by the involution θ, depending on the relative configurations of the horospheres. (See figure 2.)

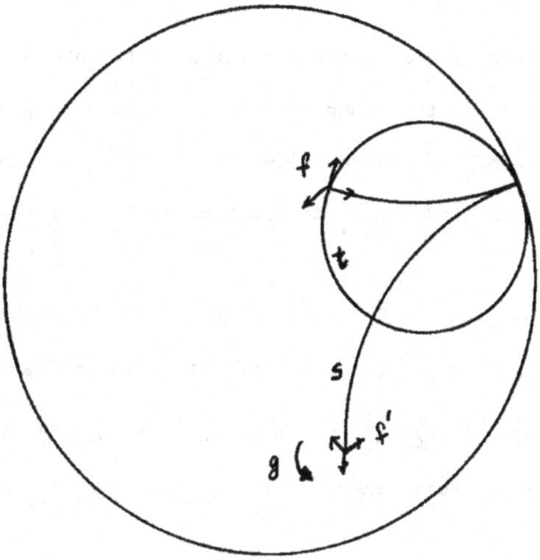

Figure 1. The Iwasawa decomposition.

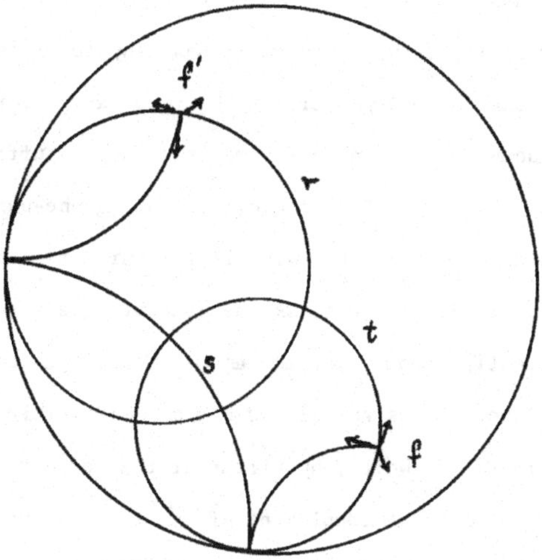

Figure 2. The Bruhat decomposition.

§5. APPLICATION TO ERGODIC THEORY

It is now possible to indicate the connection between Hedlund and Hopf's "method of asymptotic geodesics" and the group theoretic approach to geodesic flows. The link is what I have called the "Mautner lemma", although, as has been pointed out, it goes back, via von Neumann and Segal, at least as far as Poincaré and his characteristic numbers.

Theorem 5.1. Let $B, H^1, \ldots, H^m$ be C^1 vector fields which induce one parameter groups of operators $\{B_s\}$, $\{H^i_t\}$ in an appropriate Hilbert space of functions on the manifold. Suppose $[B, H^i] = \sum L^i_k H^k$, where (L^i_k) is a uniformly positive definite matrix of C^1 functions. Furthermore, let the operators $\{B_s\}$ be unitary. Then

$$\lim_{s\to\infty} B_{-s} H^i_t B_s = I$$

strongly for each t and each i.

An immediate corollary of this theorem is that any function invariant under the group $\{B_s\}$ is also fixed by every operator H^i_t. Now apply this to the case where M is compact. (∇R bounded and finite volume is actually sufficient.) Let $\mathfrak{S}$ be the set of C^1 vector fields on $\mathbb{F}$ whose corresponding one-parameter groups of operators in $L^2(\mathbb{F})$ fix every element which is Φ-invariant. Then $\mathfrak{S}$ is a submodule, over smooth functions, of $\mathfrak{X}$, and contains the bracket of any two of its members provided it is sufficiently smooth. By the above theorem, $\mathfrak{S}$ contains $\mathfrak{n}$ and $\mathfrak{n}$, and it is easy to see that $\mathfrak{S}$ is normalized by $\mathfrak{m}$. The vertical vector fields in $\mathfrak{S}$ contain a subalgebra of $\mathfrak{k}$ which, by virtue of the definite character of the matrix $L^+ - L^-$, spans at least a complement of $\mathfrak{m}$ in $\mathfrak{k}$. But such a subalgebra, normalized by $\mathfrak{m}$, must be all of $\mathfrak{k}$. It follows that $\mathfrak{S} \supset \mathfrak{X}$, and we have

Theorem 5.2. If M is an oriented compact, 1/4-pinched Riemannian manifold of negative curvature, the generalized geodesic flow $\{\Phi_t\}$ in $\mathbb{F}^+$, its oriented bundle of orthonormal frames, is ergodic.

It is easy to refine this proof slightly to obtain weak mixing. However, since $\{\Phi_t\}$ is an extension by a compact group, G_n, of the usual geodesic flow in TM, Theorem 5.2 combines with a result of Ken Thomas [16] and the known properties of the geodesic flow to give the

Corollary 5.3. $\{\Phi_t\}$ is a Kolmogorov flow in $\mathbb{F}^+$.

When the dimension of M is greater than two, the splitting (4.2) shows that $\{\Phi_t\}$ is not an Anosov flow. Rather, $G_n \times \{\Phi_t\}$ define what Hirsch, Pugh, and Shub [11] have called an Anosov action, and the ergodicity of this combined action could have been proved by appealing to Theorem (1.1) of Pugh and Shub [15]. The above result is stronger, since it tells precisely which one parameter subgroup is ergodic.

§6. PROBLEMS

At the risk of being obvious, I shall state some questions which naturally arise at this point. First, in view of the revival of interest in horocycle flows which Furstenberg's beautiful result (reported at this conference) is bound to create, we ask

1. Are the horocycle flows in manifolds of non-constant curvature minimal? strictly ergodic?

There is some "evidence by analogy" for affirmative answers to these questions. In fact, extrapolating from Moore's results in general homogeneous spaces, one can ask

2. Are the generalized horocycle flows ergodic in $\mathbb{F}$?

Aside from the dynamics, the process of drawing analogies with symmetric spaces obviously does not stop with the decompositions we have obtained. First on the agenda is the question

3. What are the relationships between the various definitions of rank for manifolds of non-positive curvature?

Here I have in mind the variations on the Chern-Kuiper (local) index of nullity, ranging up to the global assumption that $\mathfrak{p}$ in (1.3) contain an abelian subalgebra of the appropriate dimension. That a correct definition of rank might be useful to global dynamicists is suggested by some recent work of P. Eberlein, [4], where he obtains necessary and sufficient conditions that the geodesic flow on a manifold without conjugate points be Anosov. In fact, if one defines rank in terms of the number of linearly independent bounded Jacobi fields of a certain type, he has already proved a version of the conjecture

4. (Conjecture) The geodesic flow in the tangent bundle of a manifold without conjugate points is ergodic if and only if the rank is one.

Mautner, of course, obtained this result for symmetric spaces. ([14]).

In order to extend these techniques to a wider class of spaces, say, those with no conjugate points (where the horospheres still exist) or merely to those with non-positive curvature, it will be necessary to refine my analysis in order to prove smoothness. So, for technical reasons, I close with the following question, which, however, also has diagnostic applications.

5. Is the "1/4" essential in Proposition 2.3?

If in response to this question there appears in neon lights across your brow the phrase "sphere theorem", you are a differential geometer.

REFERENCES

1. Anosov, D. V., Geodesic flows on Closed Riemann manifolds with negative curvature, Trudy Steklov Inst. Mat., 90 (1967); = Translations, A.M.S., Providence, Rhode Island (1969).

2. ___________, and Sinai, Ja. G., Some smooth ergodic systems, Uspehi Mat. Nauk 22, 107-172 (1967); = Russ. Math. Surveys 22, 103-167 (1967).

3. Arnold, V. I., Some remarks on flows of line elements and frames, Dokl. Akad. Nauk SSSR 138, 255-257 (1961) = Sov. Math. Dokl. 2, 562-564 (1961).

4. Eberlein, P., When is a geodesic flow Anosov? I (to appear.)

5. ___________, and O'Neill, B., Visibility manifolds (to appear.)

6. Gelfand, I. M., and Fomin, S., Geodesic flows on manifolds of constant negative curvature, Uspehi Mat. Nauk 7, 118-137 (1952).

7. Grant, A., Surfaces of negative curvature and permanent regional transitivity, Duke Math. Jrnl. 5, 207-229 (1939).

8. Green, L. W., A theorem of E. Hopf, Michigan Math. Jrnl. 5, 31-34 (1958).

9. Hedlund, G. A., Fuchsian groups and transitive horocycles, Duke Math. Jrnl. 2, 530-542 (1936).

10. ___________, The measure of geodesic types on surfaces of negative curvature, Duke Math. Jrnl. 5, 230-248 (1939).

11. Hirsch, M. W., Pugh, C. C., and Shub, M., Invariant manifolds, Bull. Am. Math. Soc. 76, 1015-1019 (1970).

12. Hopf, E., Statistik der geodätischen Linien in Mannigfaltigkeiten negativer Krümmung, Ber. Verh. Sächs. Akad. Wiss., Leipzig 91, 261-304 (1939).

13. _______, Statistik der Losungen geodätischer Probleme vom unstabilen Typus. II, Math. Annalen 117, 590-608 (1940).

14. Mautner, F. I., Geodesic flows on symmetric Riemann spaces, Annals of Math. (2) 65, 416-431 (1957).

15. Pugh, C. C., and Shub, M., Ergodicity of Anosov actions, Inventiones Math. 15, 1-23 (1972).

16. Thomas, R. K., Metric properties of transformations of G-spaces, Trans. Am. Math. Soc. 160, 103-117 (1971).

Non-Minimality of 3-Manifolds

Otomar Hájek

Department of Mathematics and Statistics

Case Western Reserve University

This is a report on an unsuccessful attempt to prove a theorem, in the hope that other workers may be tempted to treat the problem. The conjecture, due to Gottschalk, is rather well-known:

Conjecture 1 . The 3-sphere is not minimal under any continuous flow.

The analogue is obviously false for the 1-sphere, and true for even-dimensional sheres, even for discrete flows (if $f : S^{2n} \to S^{2n}$ is continuous, then f^2 has a fixed point, [5]). I had hoped to proceed in the time-honoured fashion, by proving the more general

Conjecture 2 . [1, Problems] If X is compact, pathwise connected, and locally simply connected, and if X is minimal under a continuous flow π , then X is not simply connected.

One notes that, indeed, S^3 is simply connected, but S^1 is not. For some related results, see Chu and Geraghty [3].

The proposed proof consists of three steps: (1) To show that X has enough local sections, and that, moreover, many compact local sections can be extended appropriately (see, e.g., [6,VI,2] for precise definitions); (2) To show that there exists a local section, Y , which is inextensible; (3) To show that then $\pi : Y \times R^1 \to X$ is a covering mapping. The assertion then follows easily: by classical results on universal covering spaces, if X were simply connected, $Y \times R^1$ would have to be homeomorphic to X , although the latter is compact and the former is not. (Actually, even for non-compact X , this would show that π is a homeomorphism, even though it is minimal.)

The current state is that Steps 1 and 3 are completed, and I have reached an impasse in Step 2.

For Step 1, the needed results are provided in [6]. If X is a 2-manifold, the local sections may actually be taken to be arcs (loc. cit.; an analoguous, but incorrect, assertion is due to Whitney, [8, p.26o]). For 3-manifolds, Chewning [2] asserts that the sections may be taken as 2-manifolds. (Of course, Conjecture 2 remains interesting even if we also require that X and π be differentiable, whereupon these questions are answered trivially.)

As concerns Step 2, a proof is known if X is a 2-manifold. In the differentiable case this is a classical result of Siegel (e.g., [4,XVII,6]); the general case is somewhat more difficult, and it seems necessary to prove first that X is orientable, [7]. However, there is a stronger conclusion: the 2-torus is the only compact minimal 2-manifold.

Step 3 is rather easy: for instance, one shows that the "first intersection time with Y ", i.e., $t > 0$ such that $\pi(x,t) \in Y$ but $\pi(x,s) \notin Y$ for all $s \in (0,t)$, defines a continuous mapping $Y \to R^1$. It is for the continuity that one needs Y to be an inextensible local section.

All of this then reduces Conjecture 2 to

<u>Conjecture</u> 3 . If X is compact, and minimal under a continuius flow, then X contains a compact, connected local section which is inextensible.

References

[1] J. Auslander and W.H. Gottschalk (editors), Topological Dynamics, Benjamin, New York, 1968.

[2] W.C. Chewning, Transversal discs in R^3 , submitted to Proc. AMS.

[3] H. Chu and M.A. Geraghty, The fundamental group and the first cohomology group of a minimal set, Bull. AMS 69 (3) (1963) 377 - 381.

[4] E.A. Coddington and N. Levinson, Theory of Ordinary Differential Equations, McGraw Hill, New York, 1955.

[5] O. Hájek, Homological fixed point theorems, Comment. Math. Univ. Carol. 5 (1) (1964) 13 - 31.

[6] --- , Dynamical Systems in the Plane, Academic Press, London, 1968.

[7] --- , Lecture Notes on Dynamical Systems IX (unpublished), Case Western Reserve University, 197o.

[8] H. Whitney, Regular families of curves, Annals of Math. 34 (1933) 244-27o.

EXAMPLES OF ERGODIC MEASURE PRESERVING TRANSFORMATIONS WHICH ARE WEAKLY MIXING BUT NOT STRONGLY MIXING

Shizuo Kakutani
Yale University

1. We construct two examples of ergodic measure preserving transformations which are weakly mixing but not strongly mixing. Both examples are constructed by using the same method which can be considered as a special case of the skyscraper construction (when the building has only two floors) which was introduced as an inverse operation of taking the induced measure preserving transformation (see [1], [2]). The first of these examples was originally discussed with John von Neumann in 1941, and the second example is closely related with an example recently discussed by A. B. Katok and A. M. Stepin [3]. It is to be observed that our method is elementary and is based on a simple property of a transcendental number of Liouville type, while the arguments used by Katok and Stepin depend on some delicate properties of the continued fraction expansion of a transcendental number.

2. Let $(X, \mathcal{B}, \mu)$ be a measure space, and let φ be a one-to-one measure preserving transformation of $(X, \mathcal{B}, \mu)$ onto itself. Let A be a measurable subset of X such that $\mu(A) > 0$ and $\mu(X - A) > 0$. Let A' be a set disjoint from X and assume that there exists a one-to-one mapping τ of A onto A'. We define a measure space $(\tilde{X}, \tilde{\mathcal{B}}, \tilde{\mu})$ on $\tilde{X} = X \cup A'$ (disjoint) as follows: A subset $\tilde{B}$ of $\tilde{X}$ belongs to $\tilde{\mathcal{B}}$ if and only if $\tilde{B} \cap X \in \mathcal{B}$ and $\tau^{-1}(B \cap A') \in \mathcal{B}$, and $\tilde{\mu}(\tilde{B}) = \mu(\tilde{B} \cap X) + \mu(\tau^{-1}(\tilde{B} \cap A'))$.

Let $\tilde{\varphi}$ be a one-to-one mapping of $\tilde{X}$ onto itself defined by

$$(1) \qquad \tilde{\varphi}(\tilde{x}) = \begin{cases} \tau(\tilde{x}) & \text{if } \tilde{x} \in A, \\ \varphi(\tilde{x}) & \text{if } \tilde{x} \in X - A, \\ \varphi(\tau^{-1}(\tilde{x})) & \text{if } x \in A'. \end{cases}$$

It is easy to see that $\tilde{\varphi}$ is a one-to-one measure preserving transformation of $(\tilde{X}, \tilde{\mathcal{B}}, \tilde{\mu})$ onto itself, and that $\tilde{\varphi}$ is ergodic on $(\tilde{X}, \tilde{\mathcal{B}}, \tilde{\mu})$ if and only if φ is ergodic on $(X, \mathcal{B}, \mu)$. $\tilde{\varphi}$ is called <u>the measure preserving transformation defined on the two-storied building</u> $(\tilde{X}, \tilde{\mathcal{B}}, \tilde{\mu})$ <u>over the ground floor</u> $(X, \mathcal{B}, \mu)$ <u>by means of the ground measure preserving transformation</u> φ <u>and the base set</u> A, and is denoted by φ^A.

It is easy to see that if we start from the measure preserving transformation $\tilde{\varphi}$ defined on the measure space $(\tilde{X}, \tilde{\mathcal{B}}, \tilde{\mu})$ and consider the measure preserving transformation φ <u>induced</u> on the measure space $(X, \mathcal{B}, \mu)$ by using the method described in [2], then we obtain the original measure preserving transformation φ.

3. <u>First Example</u>. Let $(X, \mathcal{B}, \mu)$ be a measure space, where X is the set of all dyadically irrational numbers x with $0 < x < 1$, $\mathcal{B}$ is the sigma-field of all Borel subsets B of X, and $\mu(B)$ denotes the ordinary Lebesgue measure of B with the normalization $\mu(X) = 1$.

We define a one-to-one mapping φ of X onto itself by

$$(2) \qquad \varphi(x) = x - \left(1 - \frac{1}{2^n} - \frac{1}{2^{n+1}}\right) \quad \text{if } x \in I_n, \quad n = 0,1,2,\ldots,$$

where

$$(3) \qquad I_n = \left\{x \mid x \in X \text{ and } 1 - \frac{1}{2^n} < x < 1 - \frac{1}{2^{n+1}}\right\}, \quad n = 0,1,2,\ldots$$

It is easy to see that φ is an ergodic one-to-one measure preserving transformation with a pure point spectrum defined on the measure space $(X, \mathcal{B}, \mu)$ and that the complete set of eigen-values of φ is given by

(4) $$\Lambda = \left\{\lambda \,\middle|\, \lambda = 0 \text{ or } \lambda = \frac{2k-1}{2^n},\ k=1,2,\ldots,2^{n-1};\ n = 1,2,\ldots\right\}.$$

in fact, if we put $f_0(x) \equiv 1$ and

(5) $$f_\lambda(x) = \exp\left\{(2k-1)\left(\frac{\varepsilon_1}{2^n} + \frac{\varepsilon_2}{2^{n-1}} + \cdots + \frac{\varepsilon_n}{2}\right)\right.$$

if $\lambda = \frac{2k-1}{2^n}$ and if

(6) $$\frac{\varepsilon_1}{2} + \frac{\varepsilon_2}{2^2} + \cdots + \frac{\varepsilon_n}{2^n} < x < \frac{\varepsilon_1}{2} + \frac{\varepsilon_2}{2^2} + \cdots + \frac{\varepsilon_n}{2^n} + \frac{1}{2^n},$$

where $\varepsilon_i = 0$ or 1, $i = 1,2,\ldots,n$, then

(7) $$f_\lambda(\varphi(x)) = e^{2\pi i \lambda} f_\lambda(x)$$

for all $x \in X$ and that $\{f_\lambda \mid \lambda \in \Lambda\}$ is a complete orthonormal system in the complex L^2-space $L^2(X, \mathcal{B}, \mu)$. From this follows that

(8) $$\lim_{p\to\infty} \int_X \left| f(\varphi^{2^p}(x)) - f(x)\right|^2 \mu(dx) = 0$$

for any $f \in L^2(X, \mathcal{B}, \mu)$.

We now put

(9) $$A = \bigcup_{n=0}^{\infty} I_{2n} \quad \text{(disjoint)}$$

and claim that the measure preserving transformation $\tilde{\varphi} = \varphi^A$ thus defined is weakly mixing but not strongly mixing on $(\tilde{X}, \tilde{\mathcal{B}}, \tilde{\mu})$.

In order to show that $\tilde{\varphi}$ is weakly mixing on $(\tilde{X}, \tilde{\mathcal{B}}, \tilde{\mu})$, assume that there exists an eigen-function $f \in L^2(\tilde{X}, \tilde{\mathcal{B}}, \tilde{\mu})$ which belongs to an eigen-value $\lambda \not\equiv 0 \pmod 1$:

(10) $$f(\tilde{\varphi}(\tilde{x})) = e^{2\pi i \lambda} f(x)$$

for all $\tilde{x} \in \tilde{X}$. We may assume that $|f(\tilde{x})| \equiv 1$ for all $\tilde{x} \in \tilde{X}$. From the definition (1) of $\tilde{\varphi}$, it follows that

(11) $$f(\varphi(x)) = e^{2\pi i \lambda u(x)} f(x)$$

for all $x \in X$, where u is an integer-valued function defined on X by

(12) $$u(x) = \begin{cases} 2 & \text{if } x \in A, \\ 1 & \text{if } x \in X - A, \end{cases}$$

and hence

(13) $$f(\varphi^n(x)) = e^{2\pi i \lambda u_n(x)} f(x)$$

for all $x \in X$ and $n = 1,2,\ldots$, where

(14) $$u_n(x) = u(x) + u(\varphi(x)) + \ldots + u(\varphi^{n-1}(x))$$

for all $x \in X$ and $n = 1,2,\ldots$

We now observe that $u_{2^{2p}}$ is an integer-valued measurable function defined on X which takes only two different values $M_p + 1$ and $M_p + 2$, where $M_p = (5/3)(2^{2p} - 1)$, and that

(15) $$\mu\{x \mid u_{2^{2p}}(x) = M_p + 1\} = 1/3,$$

(16) $$\mu\{x \mid u_{2^{2p}}(x) = M_p + 2\} = 2/3,$$

for $p = 1,2,\ldots$ From (14), (15) and (16) follows that

(17) $$\int_X |f(\varphi^{2^{2p}}(x)) - f(x)|^2 \mu(dx) = \int_X |e^{2\pi i \lambda u_{2^{2p}}(x)} - 1|^2 \mu(dx)$$

$$= |e^{2\pi i \lambda (M_p+1)} - 1|^2 (1/3) + |e^{2\pi i \lambda (M_p+2)} - 1|^2 (2/3)$$

and, by (18), this must tend to 0 as p tends to ∞. This implies that $\lim_{p\to\infty} \lambda(M_p + 1) \equiv 0 \pmod 1$ and $\lim_{p\to\infty} \lambda(M_p + 2) \equiv 0 \pmod 1$ and hence $\lambda \equiv 0 \pmod 1$. Since $\lambda \not\equiv 0 \pmod 1$ by assumption, this is a contradiction. This shows that $\tilde{\varphi}$ is weakly mixing on $(\tilde{X}, \tilde{\mathcal{B}}, \tilde{\mu})$.

In order to show that $\tilde{\varphi}$ is not strongly mixing on $(\tilde{X}, \tilde{\mathcal{B}}, \tilde{\mu})$, we note that $\tilde{\varphi}^{u(x)}(x) = \varphi(x)$ for all $x \in X$ and hence $\tilde{\varphi}^{u_n(x)}(x) = \varphi^n(x)$ for all $x \in X$ and $n = 1,2,\dots$ If we put $n = 2^{2p}$, it follows that $\tilde{\varphi}^{u_{2^{2p}}(x)}(x) = \varphi^{2^{2p}}(x)$ and hence $\tilde{\varphi}^{M_p(x)} = \tilde{\varphi}^{-1}(\varphi^{2^{2p}}(x))$ or $\tilde{\varphi}^{M_p(x)} = \tilde{\varphi}^{-2}(\varphi^{2^{2p}}(x))$ according as $u_{2^{2p}}(x) = M_p + 1$ or $u_{2^{2p}}(x) = M_p + 2$. From this follows that

$$(18) \qquad \tilde{\varphi}^{M_p}(B) \subset \tilde{\varphi}^{-1}(\varphi^{2^{2p}}(B)) \cup \tilde{\varphi}^{-2}(\varphi^{2^{2p}}(B))$$

for any $B \in \mathcal{B}$ and $p = 1,2,\dots$ Let us put $B = I_k$ for some k $(k = 0,1,2,\dots)$. Then it is easy to see that $\varphi^{2^{2p}}(B) = B$ for $2p \geqq k$ and this shows that $\tilde{\varphi}$ is not strongly mixing on $(\tilde{X}, \tilde{\mathcal{B}}, \tilde{\mu})$.

4. Second Example. Let $(X, \mathcal{B}, \mu)$ be a measure space, where X is the set of all real numbers mod 1, $\mathcal{B}$ is the sigma-field of all Borel subsets B of X, and $\mu(B)$ is the ordinary Lebesgue measure of B with the normalization $\mu(X) = 1$. We define a one-to-one mapping φ of X onto itself by

$$(19) \qquad \varphi(x) \equiv x + \alpha \pmod 1,$$

where α is a transcendental number of Liouville type defined by

$$(20) \qquad \alpha = \sum_{k=1}^{\infty} 10^{-n_k},$$

where $\{n_k \mid k = 1,2,\dots\}$ is an increasing sequence of positive integers such that

$$(21) \qquad \lim_{k\to\infty} (n_{k+1} - 2n_k) = +\infty.$$

φ is obviously an ergodic measure preserving transformation with a pure point spectrum defined on the measure space $(X, \mathcal{B}, \mu)$, and it is easy to see that

$$(22) \qquad \lim_{k\to\infty} \int_X |f(\varphi^{10^{n_k}}(x)) - f(x)|^2 \mu(dx) = 0$$

for any $f \in L^2(X, \mathcal{B}, \mu)$. This follows from the fact that the complete set of eigen-values of φ is given by

$$(23) \qquad \Lambda = \{\lambda = n\alpha \mid n = 0, \pm 1, \pm 2, \ldots\}$$

and that $\lim_{k\to\infty} 10^{n_k}\alpha \equiv 0 \pmod 1$.

Let β be a real number of the form:

$$(24) \qquad \beta = \sum_{n=1}^{\infty} b_n 10^{-n} ,$$

where $b_n = 0,1,2,\ldots,8$ or 9 for $n = 1,2,\ldots$ and $b_{n_k+1} = 5$ for $k = 1,2,\ldots$ (It is to be observed that β has no need to be an irrational number. For example, if $b_n = 5$ for $n = 1,2,\ldots$, then $\beta = 5/9$ and this β will do). Let A be a subset of X defined by

$$(25) \qquad A = \{x \mid 0 \leqq x < \beta\},$$

and let us consider the measure preserving transformation $\tilde{\varphi} = \varphi^A$. We claim that $\tilde{\varphi}$ is again weakly mixing but not strongly mixing on the measure space $(\tilde{X}, \tilde{\mathcal{B}}, \tilde{\mu})$.

The proof of this statement can be carried out in essentially the same way as in the first example by using the relation (22) and by observing the fact that, if we define the functions u and u_n exactly the same way as in above, then $u_{10^{n_k}}$ is an integer-valued function which takes only two different values N_k and $N_k + 1$, where $N_k = 10^{n_k}(1 + \sum_{n=1}^{n_k} b_n 10^{-n})$, and that

$$(26) \qquad \mu\{x \mid u_{10^{n_k}}(x) = N_k\} \geqq 3/10,$$

$$(27) \qquad \mu\{x \mid u_{10^{n_k}}(x) = N_k + 1\} \geqq 4/10,$$

for sufficiently large k. (We observe that $u_{10^{n_k}}$ is an integer-valued step function defined on the set X of real numbers mod 1 whose discontinuities appear only at $\gamma_j \equiv -j\alpha$ (mod 1) and $\delta_j \equiv -j\alpha + \beta$ (mod 1) for $j = 0,1,\ldots, 10^{n_k} - 1$, and that the amount of jump $u_{10^{n_k}}(x+0) - u_{10^{n_k}}(x-0)$ at each γ_j and δ_j is $+1$ and -1, respectively, and further that these γ_j's and δ_j's are located alternatively on the set X of real numbers mod 1 in such a way that each interval $[(i-1)\cdot 10^{-n_k} + 9\cdot 10^{-n_k-1}, i\cdot 10^{-n_k}]$ contains exactly one of the γ_j's and each interval $[i\,10^{-n_k} + 4\cdot 10^{-n_k-1}, i\cdot 10^{-n_k} + 6\cdot 10^{-n_k-1}]$ contains exactly one of the δ_j's for $i = 1,2,\ldots, 10^{n_k}$).

Bibliography

1. Hajian, A. B. and Kakutani, S., Example of an ergodic measure preserving transformation on an infinite measure space, Contributions to Ergodic Theory and Probability, Proceedings of the First Midwestern Conference on Ergodic Theory held at Ohio State University, March 27-30, 1970; Lecture Notes in Mathematics, No. 170, pp. 45-52.
2. Kakutani, S., Induced measure preserving transformations, Proc. Acad. Tokyo 19(1943), pp. 635-641.
3. Katok, A. B. and Stepin, A. M., Approximations in ergodic theory, Uspehi Matematicheskii Nauk 22(1967), pp. 81-106.

LOCALLY CONNECTED ALMOST PERIODIC MINIMAL SETS

Joseph F. Kent

1. Introduction. Using properties of almost periodic functions and their spectra, M. L. Cartwright [3] has studied compact almost periodic orbit closures in R^n, Euclidean n-space. Some of her results provide necessary and sufficient conditions for such orbit closures to be homeomorphic to tori. Topological group results [9, page 262] imply that every compact locally connected almost periodic minimal set in a second countably space is homeomorphic to a torus. None of these results answers the following natural question: in such a case, is the almost periodic minimal flow isomorphic to an irrational flow on the torus? The main aim of this paper is to answer this question in the affirmative for locally connected almost periodic minimal sets in R^n on C^n, complex n-space. By showing that the character group of an almost periodic minimal set X in R^n or C^n is algebraically isomorphic to the subgroup of R generated by the associated spectrum S(X), we are able to make use of both character group and spectrum results. Locally connected almost periodic minimal sets are shown to be naturally associated with quasiperiodic functions [7] which occur in the study of nonlinear oscillations [6].

Section 2 is devoted to preliminary definitions and remarks. In section 3 character groups, almost periodic functions, and spectra are introduced, and the relationships among them are examined. An irrational flow on a torus is examined in section 4 in the light of the concepts previously introduced. Section 5 contains the main theorem and its corollaries. Numbers in square brackets refer to item in the bibliography. The author would like to thank Professor J. W. England who first introduced him to almost periodic minimal flows.

2. Almost periodic flows. Throughout the paper X will denote a compact subset of R^n or C^n. A flow [4,5,8] on X is a continuous function $\emptyset:X\times R \to X$ with the property that $\emptyset(x,o) = x$ and $\emptyset(\emptyset(x,t),s) = \emptyset(x,t+s)$ for all $x \in X$ and all $t,s \in R$. The orbit of a point x of X under the flow $\emptyset$ is the set of all points $\emptyset(x,t)$, $t \in R$ and is denoted by $0(x)$. A closed subset A of X is said to be minimal if $cl[0(x)] = A$ for all x in A, where $cl[0(x)]$ denotes the closure in X of $0(x)$. Two flows $\emptyset,\Psi$ defined on X,Y, respectively, are said to be (flow) isomorphic if there is a homeomorphism $h:X \to Y$ with $h(\emptyset(x,t)) = \Psi(h(x),t)$ for all t in R, all x in X.

A flow $\emptyset$ on X is said to be (uniformly) almost periodic provided that for any positive number ϵ there is a relatively dense subset A of R with $\|\emptyset(x,t) - \emptyset(x,t+s)\| < \epsilon$ for all $s \in A$, $x \in X$, and $t \in R$. Here $\|\cdot\|$ denotes the usual norm function on R^n or C^n, and a subset A of R is relatively dense if there is a positive number L such that every interval of length L has non-empty intersection with A. Since X is a complete metric space, if $\emptyset$ is an almost periodic flow on X, then X is a compact minimal set. The irrational flow on a torus discussed in section 4 is an example of an almost periodic flow. The following theorem is a well known classical result in the study of such minimal sets.

Theorem 2.1 [8, page 394; 5, page 39]: Let X be a compact subset of a complete metric space. The following are equivalent:

(i) X admits an almost periodic flow ϕ;

(ii) For any x in X, X admits a topological group structure such that x is the identity, and there is a continuous homomorphism $h: R \to X$ onto a dense subgroup;

(iii) X is the space of a compact solenoidal topological group.

The homomorphism h in (ii) induces the flow ϕ in (i) such that $\phi x = h$ where $\phi x(t) = \phi(x,t)$. Conversely, given the flow ϕ, the homomorphism in (ii) is ϕx.

Recall that a compact topological group G is solenoidal provided there is a continuous homomorphism $h: R \to G$ onto a dense subgroup. Any such group is connected and abelian. When considering such a group G we will occasionally write (G,h) in order to specify the homomorphism. If (G,h) and (H,k) are compact solenoidal topological groups in a metric space and there is an isomorphism $F: G \to H$ with $k = F \circ h$, then the associated almost periodic minimal flows are isomorphic via F. On the other hand (G,h) and (H,k) may admit a topological group isomorphism when the associated almost periodic flows are not isomorphic. An example of this is noted in section 4.

3. Character groups, almost periodic functions, and spectra. If G is a topological group, then G^* will denote the group of continuous characters of G [9]. If $f: G \to H$ is a homomorphism of topological groups, then $f^*: H^* \to G^*$ will denote the induced map given by

$f^*(h^*) = h^* \circ f$ for h^* in H^*. For algebraic and topological relationships between G and G*, the reader is referred to [9]. We will only need to apply the theory to finite dimensional compact solenoidal topological groups.

Recall that the rank of an abelian group G is an invariant equal to the cardinality of a maximal linearly independent subset of G. A subset H of G is linearly independent if whenever $\{g(1),g(2),\ldots,g(n)\}$ is a finite set of elements of H and $a(1),\ldots,a(n)$ are integers with $a(1)g(1) + a(2)g(2) + \ldots + a(n)g(n) = 0$, then $a(i) = 0$ for all $i = 1,\ldots,n$. Such a linearly independent set H is maximal if $\{x\} \cup H$ is a linearly dependet set for any x in $G\backslash H$, the compliment of H in G. For a finitely generated abelian group, the rank is the minimal number of generators of the torsion-free part of the canonical decomposition.

Remark 3.1: If G is a compact abelian group, then the dimension of G is the rank of G*.

Remark 3.2: If G is a compact finite dimensional solenoidal topological group, then G* is a discrete torsion-free abelian group of finite rank, and G* is finitely generated if and only if G is locally connected.

Remark 3.3 [5, page 39]: Let G be a compact abelian group. G is solenoidal if and only if G* is algebraically isomorphic to a subgroup of R. In fact, if (G,h) is a compact solenoidal topological group, then $h^*:G^* \to h^*(G^*) \subseteq R^* = R$ is an isomorphism. Conversely, if H is a subgroup of R, and α is the inclusion homomorphism, then (H^*,α^*) is a compact solenoidal topological group.

A continuous function $f:R \to R^n$ or $f:R \to C^n$ is said to be almost periodic [2,4] provided that given $\epsilon > 0$, there is a relatively dense subset A of R with $\|f(t+s) - f(t)\| < \epsilon$ for all s in A, t in R. A vector-valued ($n > 1$) continuous function is almost periodic if and only if all of its component functions are scalar-valued almost periodic functions. An almost periodic function f defines a flow ϕ on f(R) by $\phi(f(t),s) = f(t+s)$, and ϕ may be extended uniquely to an almost periodic flow on cl[f(R)] [3]. Conversely, if X is an almost periodic minimal set under the flow ϕ, then ϕx is an almost periodic function for any x in X.

For an almost periodic function f, we define the mean of f, denoted M(f(t)), to be the limit as T approaches infinity of $(1/T)\int_0^T f(t)dt$ provided the limit exists. Otherwise, M(f(t)) is undefined. (Here the integral is the usual Riemann integral.) For any almost periodic function f and real number λ, $f(t)\exp(-i\lambda t)$ is also an almost periodic function. $M(f(t)\exp(-i\lambda t))$ exists for all real λ, and in fact, $M(f(t)\exp(-i\lambda t))$ is not the zero vector for at most a countable set of λ's. The spectrum of f is denoted S(f) and defined to be the set of values of λ for which $M(f(t)\exp(-i\lambda t))$ is not the zero vector. The spectrum is used to provide a Fourier analysis of f [2].

A subset H of the reals or of any rational vector space is said to be rationally independent if whenever $\{h(1),h(2),\ldots,h(n)\}$ is a finite set of elements of H, and $r(1),r(2),\ldots,r(n)$ are rational numbers with $r(1)h(1) + r(2)h(2) + \ldots + r(n)h(n) = 0$, we have $r(i) = 0$ for all $i = 1,\ldots,n$. Considering a torsion-free abelian group G as a subset of the rational vector space $G\otimes Q$, where Q is the set of rational numbers, we see that linear independence and rational independence coincide.

Let $S(f)$ be the spectrum of the almost periodic function f. If there exist rationally independent real numbers $\omega(1),\omega(2),\ldots$ with the property that for each $\lambda \in S(f)$, there are rational numbers $r(\lambda,j)$ for $j = 1,\ldots,N(\lambda)$ with $\lambda = \Sigma\, r(\lambda,j)\,\omega(j)$, then we say $\{\omega(1),\omega(2),\ldots\}$ is a rational base for $S(f)$. If for all $\lambda \in S(f)$ and all $j = 1,\ldots,N(\lambda)$, $r(\lambda,j)$ is an integer, then $\{\omega(1),\omega(2),\ldots\}$ is called an integral base. If the supremum of the integers $N(\lambda)$, taken over all λ in $S(f)$, is finite, then we have a finite base. An almost periodic function with a finite integral base is said to be quasiperiodic [2,7]. Moser [7] uses the following characterization of such functions, which follows from Besicovitch [2, page 37] and Fourier analysis of almost periodic functions.

Remark 3.4: f is a (vector-valued) almost periodic function with a finite integral base $\{2\pi\omega(1),2\pi\omega(2),\ldots,2\pi\omega(n)\}$ provided there is a continuous (vector-valued) function $F(x(1),x(2),\ldots,x(n))$ with period 1 in each variable $x(i)$ such that $f(t) = F(t\omega(1), t\omega(2),\ldots,t\omega(n))$.

If X is an almost periodic minimal set under the flow $\emptyset$, then for any $x \in X$ we may associate with x the set $S(\emptyset x)$. Clearly for $y \in O(x)$, $S(\emptyset y) = S(\emptyset x)$. For $y \notin O(x)$, there is a sequence $y(1),y(2),\ldots$ of points of $O(x)$ such that $\emptyset y(1),\emptyset y(2),\ldots$ converges uniformly to $\emptyset y$. From this it easily follows that for any x,y in X, $S(\emptyset x) = S(\emptyset y)$, and we denote this set by $S(X)$. Our first result is to connect the character group of X with the subgroup of the reals generated by $S(X)$, denoted by $\langle S(X)\rangle$.

Theorem 3.5: Let X be a compact almost periodic minimal set in R^n or C^n under the flow ϕ. Then for any $x \in X$ $\phi x^*(X^*) = \langle S(X) \rangle$.

This result is easily seen to be equivalent to: if f is a vector-valued almost periodic function, then $\langle S(f) \rangle = f^*(cl[f(R)]^*)$ This type of result was first proven by Auslander and Hahn [1] for scalar-valued almost periodic functions. The following two lemmas when f is scalar-valued are taken from their paper [1, page 136].

Lemma 3.6: Let f be an almost periodic function, and let H be a subgroup of the additive group of reals with the discrete topology. If $g = H^*$ and $\alpha : R \to G$ is the dual of the inclusion $\alpha^* : H \to R$, then there is a continuous function F on G with $f = F \circ \alpha$.

Proof: As noted the lemma has been proven for scalar-valued functions. Let $f = (f_1, \ldots, f_n)$ be a vector-valued almost periodic function with scalar-valued component functions f_i, $i = 1, \ldots, n$. For each $i = 1, \ldots, n$ we have a continuous scalar-valued function F_i with $f_i = F_i \circ \alpha$. Let $F = (F_1, \ldots, F_n)$, and the result is proven.

Lemma 3.7: Let f be an almost periodic function, and let (G, α) be a compact solenoidal topological group. If there is a continuous function F on G with $f = F \circ \alpha$, then $\langle S(f) \rangle \subseteq \alpha^*(G^*)$.

Proof: Once again from the validity of the scalar-valued case, we can conclude for a vector-valued almost periodic function $f = (f_1, \ldots, f_n)$ that $\langle S(f_i) \rangle \subseteq \alpha^*(G^*)$. An elementary argument shows that $S(f)$ is the union of $S(f_i)$ taken over $i = 1, \ldots, n$. As $S(f_i) \subseteq \langle S(f_i) \rangle \subseteq \alpha^*(G^*)$ for all i, we know $S(f) \subseteq \alpha^*(G^*)$. Since $\alpha^*(G^*)$ is a subgroup, $\langle S(f) \rangle \subseteq \alpha^*(G^*)$.

Proof of Theorem 3.5: Let X be a compact almost periodic minimal set under the flow $\emptyset$. For x in X, let $f = \emptyset x$. Then f is a vector-valued almost periodic function and $(X,\emptyset x) = (cl[f(R)],f)$ is a compact solenoidal topological group. Apply Lemma 3.7 with $\alpha = f$ and F as the identity map to conclude $\langle S(X)\rangle = \langle S(\emptyset x)\rangle = \langle S(f)\rangle \subseteq f^*(cl[f(R)]^*) = \emptyset x^*(X^*)$.

Now letting $H = \langle S(X)\rangle$ we apply Lemma 3.6 and obtain a continuous function F on $H^* = \langle S(X)\rangle^* = \langle S(f)\rangle^*$ with $f = F\circ\alpha$. F is an onto homomorphism of compact solenoidal topological groups $(\langle S(f)\rangle^*,\alpha)$ and $(cl[f(R)],f)$ with $f = F\circ\alpha$, and hence F induces the containment $f^*(cl[f(R)]^*) \subseteq \alpha^*(\langle S(f)\rangle^{**}) = \langle S(f)\rangle$. This completes the proof.

If f is an almost periodic function, then a finite rational (integral) base $\{\omega(1),\ldots,\omega(n)\}$ is said to be minimal if the rational vector spaces (abelian groups) generated by $\{\omega(1),\ldots,\omega(n)\}$ and S(f) coincide. Any two finite minimal rational (integral) bases have the same cardinality.

Corollary 3.8: If X is a compact almost periodic minimal set under a flow $\emptyset$, then X has dimension m if and only if S(X) has a minimal rational base of cardinality m.

M. L. Cartwright [3, Theorem 8] has proven the sufficiency of this result by selecting a particular minimal rational base and using the spatial extension of an almost periodic function. We will provide an alternate proof via Theorem 3.5 and Remark 3.1.

Proof of Corollary 3.8: Assume X has dimension m. Fix x in X to be identity element of the solenoidal topological group $(X,\emptyset x)$. By Remark 3.1 X^* has rank m. As $\emptyset x^*$ is a monomorphism $\emptyset x^*(X^*)$ has rank m. Therefore by the theorem $\langle S(X)\rangle$ has rank m. Thus $\langle S(X)\rangle$ has a maximal linearly independent (and hence rationally independent) set $\{\omega(1),\ldots,\omega(m)\}$. For any $s \in \langle S(X)\rangle\backslash\{\omega(1),\ldots,\omega(m)\}$, we know by maximality that there are integers $p(s,0)$, $p(s,1),\ldots, p(s,m)$, not all zero, with $p(s,0)s + p(s,1)\omega(1) + \ldots + p(s,m)\omega(m) = 0$. Therefore by the linear independence of the $\omega(i)$ we know $p(s,0) \neq 0$, and $s = r(s,1)\omega(1) + \ldots + r(s,m)\omega(m)$ where $r(s,i)$ is the rational number $-p(s,i)\backslash P(s,0)$. We see $\{\omega(1),\ldots,\omega(m)\}$ is a finite rational base for $S(X)$, and it is minimal as it is a basis for $\langle S(X)\rangle$ considered as a rational vector-space.

Conversely, if $S(X)$ has a minimal rational base $\{\omega(1),\ldots, \omega(m)\}$ of cardinality m, we see that $\langle S(X)\rangle$ is the rational vector space generated by the $\omega(i)$, $i = 1,\ldots,m$. Any element s of $\langle S(X)\rangle\backslash\{\omega(1),\ldots,\omega(m)\}$ is of the form $r(s,1)\omega(1) + \ldots + r(s,m)\omega(m)$ for unique rationals $r(s,i)$, $i = 1,\ldots,m$. This implies $\{s,\omega(1),\ldots,\omega(m)\}$ is a linearly dependent set. As the $\omega(i)$ are rationally independent, they are linearly independent and form a maximal linearly independent set. Thus $\langle S(X)\rangle$ has rank m, along with $\emptyset x^*(X^*)$ and X^*. Using Remark 3.1 X has dimension m.

Corollary 3.9: If X is a compact almost periodic minimal set under a flow $\emptyset$, then $S(X)$ has a finite integral base if and only if X is locally connected.

Proof: Fix $x \in X$ and assume $S(X)$ has a finite integral base $\{2\pi\omega(1),\ldots,2\pi\omega(m)\}$. As $S(X) = S(\emptyset x)$, $\emptyset x$ is a quasiperiodic function. Remark 3.4 implies there is a continuous function $F: R^m/Z^m \to X$ with $F(t\omega(1),\ldots,t\omega(m)) = \emptyset x(t)$. Here R^m/Z^m is the quotient of the additive group R^m and the subgroup Z^m of integer lattice points.

We refer to this as the additive torus. Let $\Psi:(R^m/Z^m)\times R \to R^m/Z^m$ be the irrational flow on R^m/Z^m given by $\Psi(x(1),\ldots,x(m)) = (x(1) + t\omega(1),\ldots,x(m) + t\omega(m))$, modulo one in each coordinate. R^m/Z^m is an almost periodic minimal set, and if o is the zero element of R^m/Z^m we have $F\circ\Psi o = \emptyset x$. F is an onto homomorphism of compact solenoidal topological groups, and therefore F* is a monomorphism of X* into $(R^m/Z^m)^*$. $\Psi o(R)$ is a dense subgroup of not only $(R^m/Z^m,\Psi o)$ but also R^m/Z^m with the usual group structure. This is due to the fact that the group operation in $\Psi o(R)$ is simply componentwise addition modulo one. Using a result of Pontryagin [8, page 274] we see that the character groups of $(R^m/Z^m,\Psi o)$ and R^m/Z^m are isomorphic. As $(R^m/Z^m)^*$ is finitely generated with m generators, we know X* is finitely generated. By Remark 3.2 X is locally connected.

Conversely, if X is locally connected and $x \in X$, then X* is finitely generated. As $\emptyset x^*(X^*) = \langle S(X)\rangle$, $\langle S(X)\rangle$ has a finite set of rationally independent generators $\{\omega(1),\ldots,\omega(m)\}$. Since S(X) is a subset of $\langle S(X)\rangle$, for every $s \in S(X)$ there are integers $p(s,i)$, $i = 1,\ldots,m$ with $s = p(s,1)\omega(1) + \ldots + p(s,m)\omega(m)$. Hence S(X) has a finite integral base.

Corollary 3.10: Let f be an almost periodic function. f is a quasiperiodic function if and only if cl[f(R)] is locally connected.

4. Irrational flow on a torus. Let T^n denote the set of vectors in C^n with all coordinates having absolute value 1. T^n is the (multiplicative) n-torus and is isomorphic as a compact topological group with R^n/Z^n. For convenience of notation we will assume in the following example that n = 2.

A flow $\Psi: T^2 \times R \to T^2$ given by $\Psi(z(1),z(2),t) = (z(1)\exp(2\pi it\omega(1)), z(2)\exp(2\pi it\omega(2)))$, where $\omega(1)$ and $\omega(2)$ are rationally independent real numbers, is called an irrational flow on the torus. (The terminology is derived from the classical special case where $\omega(1)$ is rational and $\omega(2)$ is irrational.) The flow Ψ is isomorphic with the flow $\emptyset$ defined on R^2/Z^2 by $\emptyset(x(1),x(2),t) = (x(1) + t\omega(1), x(2) + t\omega(2))$. Both give examples of locally connected almost periodic minimal sets.

Remark 4.1: If Ψ is the irrational flow on T^2 determined by $\{\omega(1),\omega(2)\}$, then $S(T^2) = \{2\pi\omega(1), 2\pi\omega(2)\}$. If $f = \Psi(1,1)$, then $f^*(T^{2*}) = \{2\pi n\omega(1) + 2\pi m\omega(2): n,m \text{ integers}\} = \langle S(T^2)\rangle$.

The remark provides an opportunity to compute the spectrum, character group, and verify Theorem 3.5 in this case. $S(T^2) = S(f) = \{\lambda \in R: M(f(t)\exp(-i\lambda t)) \neq 0\}$, and we note that $M(f(t)\exp(-i\lambda t)) = (M(\exp(2\pi it\omega(1))\exp(-i\lambda t)), M(\exp(2\pi it\omega(2))\exp(-i\lambda t))$ $= (M(\exp(it(2\pi\omega(1) - \lambda))), M(\exp(it(2\pi\omega(2) - \lambda))))$. Hence $M(f(t)\exp(-i\lambda t))$ is not the zero vector if and only if either $2\pi\omega(1) = \lambda$ or $2\pi\omega(2) = \lambda$. Therefore $S(f) = \{2\pi\omega(1), 2\pi\omega(2)\}$.

The important thing to notice about $(T^2, \Psi x)$, considered as a solenoidal topological group with identity x, is that its character group may be identified with the character group of the dense subgroup $\Psi x(R)$. This is due to the following result [8, page 274]: if H is a subgroup of a topological group G, then $G^*/A(G,H)$ is identified with H^*, where $A(G,H)$ is the set of characters of G that contain H in their kernel. $G^*/A(G,H)$ is identified with the set of characters which are restrictions to H of characters of G. Hence every character of H is the restriction to H of a character of G. If H is dense in G, then A(G,H) is trivial and G^*, H^* are identified as above. Now $\Psi x(R)$ is a dense subgroup of T^2 considered as the product of two circle groups, since the operation in $\Psi x(R)$ induced by Ψx is componentwise multiplication. Therefore we identify naturally the character group of $(T^2, \Psi x)$ with the usual character group of T^2 which is $\{\gamma(n,m): n,m$ are integers, $\gamma(n,m)(z(1),z(2)) = z(1)^n z(2)^m\}$. Now $f^*: T^2{}^* \to R^*$ where R^* is identified with R by $s^* \to s$ where $s^*(t) = \exp(ist)$ for $s \in R$. Therefore $f^*(T^2{}^*) = \{f^*(\gamma(n,m)): n,m$ integers$\}$. Note that $f^*(\gamma(n,m))(t) = \gamma(n,m)(f(t)) = \gamma(n,m)(\exp(2\pi it\omega(1)), \exp(2\pi it\omega(2))) = \exp(2\pi nit\omega(1))\exp(2\pi mit\omega(2)) = \exp(i(2\pi n\omega(1) + 2\pi m\omega(2))t)$. Under the identification of R^* with R, we have $f^*(T^2{}^*) = \{2\pi n\omega(1) + 2\pi m\omega(2): n,m$ integers$\} = \langle S(T^2)\rangle$.

The following theorem sums up and extends the work on the previous example.

Theorem 4.2: Let $\{\omega(1),\ldots,\omega(n)\}$, $\{\alpha(1),\ldots,\alpha(n)\}$ be two sets of rationally independent real numbers. Let f,g be functions from R into T^n given by $f(t) = (\exp(2\pi it\omega(1)),\ldots,\exp(2\pi it\omega(n)))$ and $g(t) = (\exp(2\pi it\alpha(1)),\ldots,\exp(2\pi it\alpha(n)))$. If M(f) and M(g) denote the compact almost periodic minimal sets with flows $\Psi, \emptyset$ induced by f,g, respectively, then

(i) M(f), M(g), and T^n are isomorphic topological groups;

(ii) $f^*(M(f)^*) = \langle S(f)\rangle = \langle 2\pi\omega(1),\ldots,2\pi\omega(n)\rangle$ and $g^*(M(g)^*) = \langle S(g)\rangle = \langle 2\pi\alpha(1),\ldots,2\pi\alpha(n)\rangle$;

(iii) $\langle S(f)\rangle = \langle S(g)\rangle$ if and only if there exists an nxn matrix N with integer entries and determinate ± 1 such that $N(\omega(1),\ldots,\omega(n)) = (\alpha(1),\ldots,\alpha(n))$. (Note that the order of the $\omega(i)$ and $\alpha(i)$ is irrelevant as any permutation matrix has determinate ± 1 and only zeros and ones as entries.);

(iv) $\langle S(f)\rangle = \langle S(g)\rangle$ if and only if $M(f)$ and $M(g)$ are flow isomorphic.

Proof: (i) The discussion in the proof of the preceeding remark indicates that the character groups of $M(f)$, $M(g)$, and T^n may be identified. By the fundamental duality theorem of compact topological groups [8], $M(f)$, $M(g)$, and T^n are isomorphic.

Part (ii) is immediate from the argument of the preceeding remark.

(iii) Suppose there exists an $n \times n$ matrix N whose i,jth entry is the integer $n(i,j)$ and $\det(N) = \pm 1$. Then N^{-1} is a matrix with $\det(N^{-1}) = \pm 1$, and whose i,jth entry is an integer $m(i,j)$. As $N(\omega(1),\ldots,\omega(n)) = (\alpha(1),\ldots,\alpha(n))$ we have for all $i = 1,\ldots,n$, $2\pi\alpha(i) = 2\pi n(i,1)\omega(1) + \ldots + 2\pi n(i,n)\omega(n)$. Thus $2\pi\alpha(i) \in \langle S(f)\rangle$ for all i, which implies $\langle S(g)\rangle \subseteq \langle S(f)\rangle$. Using N^{-1} we have $2\pi\omega(i) = 2\pi m(i,1)\alpha(1) + \ldots + 2\pi m(i,n)\alpha(n)$, and $\langle S(f)\rangle \subseteq \langle S(g)\rangle$. Therefore $\langle S(f)\rangle = \langle S(g)\rangle$.

Conversely, if $\langle S(f)\rangle = \langle S(g)\rangle$, then we may conclude that $\langle \omega(1),\ldots,\omega(n)\rangle = \langle \alpha(1),\ldots,\alpha(n)\rangle$. Therefore there are integers $m(i,j), n(k,i)$ with $\omega(i) = m(i,1)\alpha(1) + \ldots + m(i,n)\alpha(n)$, and $\alpha(k) = n(k,1)\omega(1) + \ldots + n(k,n)\omega(n)$. Hence for $k = 1,\ldots,n$

$$\alpha(k) = \sum_{i=1}^{n} \sum_{j=1}^{n} n(k,i)m(i,j)\alpha(j) = \sum_{j=1}^{n} \left[\sum_{i=1}^{n} n(k,i)m(i,j)\right]\alpha(j).$$

Rational independence implies that $\sum_{i=1}^{n} n(k,i)m(i,j) = 1$ if $k = j$, and 0 otherwise. Let M be the $n \times n$ matrix whose i,jth entry is $m(i,j)$, and let N be the $n \times n$ matrix whose k,ith entry is $n(k,i)$.

Then NM is the identity matrix, and $\det(N)\det(M) = 1$. As both M,N have integer valued determinates, we must have $\det(N) = \pm 1$, and $\det(M) = \pm 1$. Noting that $N(\omega(1),\ldots,\omega(n)) = (\alpha(1),\ldots,\alpha(n))$, we complete part (iii) of the theorem.

(iv) If M(f) and M(g) are flow isomorphic via a map $h: M(f) \to M(g)$, then $h \circ f = g'$ and $h^{-1} \circ g' = f$, where $g' = \emptyset h(f(0))$. Now $g = \emptyset g(0)$ implies $S(g) = S(g')$. The maps h and h^{-1} induce the inclusions $g'^*(M(g')^*) \subseteq f^*(M(f)^*)$, $f^*(M(f)^*) \subseteq g'^*(M(g')^*)$, respectively. Therefore by Theorem 3.5 $\langle S(f)\rangle = \langle S(g')\rangle$, and $S(g) = S(g')$ gives the result.

Now suppose $\langle S(f)\rangle = \langle S(g)\rangle$. Then by part (iii) there is a matrix N with $N(\omega(1),\ldots,\omega(n)) = (\alpha(1),\ldots,\alpha(n))$. As $\det(N) = \pm 1$ and N has integer entries, N induces an automorphism N′ of the additive torus R^n/Z^n. Let H denote the isomorphism of R^n/Z^n and T^n given by $H(x_1,\ldots,x_n) = (\exp(2\pi i x_1),\ldots,\exp(2\pi i x_n))$. Then $H \circ N' \circ H^{-1} = H \circ N \circ H$ is a well defined automorphism of T^n. As $N(\omega(1),\ldots,\omega(n)) = (\alpha(1),\ldots,\alpha(n))$, we easily check that $(H \circ N \circ H^{-1}) \circ f = g$. Thus M(f) is flow isomorphic to M(g) via the mapping $H \circ N \circ H^{-1}$. This completes the proof of the theorem.

5. An isomorphism theorem. If X is a locally connected almost periodic minimal set in R^n or C^n, then as the space is second countable we know [8, page 262] that X must be isomorphic as a topological group to T^m for some $m < n$. Similarly, as any such X has the property that S(X) has a finite integral base of cardinality m, the corollary to Theorem 9 of [3] will imply X is homeomorphic to T^m. We may now prove the following:

Theorem 5.1: Any locally connected almost periodic minimal set in R^n or C^n is flow isomorphic to a torus with an irrational flow.

Proof: Clearly it suffices to prove that if f is a vector-valued quaisperiodic function, then cl[f(R)], considered as a locally connected almost periodic minimal set, is flow isomorphic to a torus with an irrational flow.

Let f be a quasiperiodic function with a minimal integral base $\{2\pi\omega(1),\ldots,2\pi\omega(m)\}$. Let $\Psi:(R^m/Z^m)\times R \to R^m/Z^m$ be the irrational flow given by $\Psi(x(1),\ldots,x(m),t) = (x(1) + t\omega(1),\ldots,x(m) + t\omega(m))$. Remark 3.4 implies there is a continuous function F on R^m/Z^m with $F\circ\Psi 0 = f$, where 0 is the zero of R^m/Z^m. As $\{2\pi\omega(1),\ldots,2\pi\omega(m)\}$ is a minimal integral base, we have $\langle S(f)\rangle = \langle 2\pi\omega(1),\ldots,2\pi\omega(m)\rangle = \langle S(\Psi 0)\rangle$. The relation $F\circ\Psi 0 = f$ implies $\langle S(f)\rangle = f^*(cl[f(R)]^*) = \Psi 0^*\circ F^*(cl[f(R)]^*)$. Since $\Psi 0^*$ is a monomorphism, we see that if F^* fails to be onto, then $\Psi 0^*(F^*(cl[f(R)]^*)) \neq \Psi 0^*((R^m/Z^m)^*) = \langle S(\Psi 0)\rangle$. However this contradicts $\langle S(f)\rangle = \langle S(\Psi 0)\rangle$, and so F^* is onto. This implies F is a one-to-one map and hence a homeomorphism. The relation $f = F\circ\Psi 0$ implies F is a flow isomorphism.

Two corollaries follow immediately from the theorem.

Corollary 5.2: If f,g are almost periodic functions with $\langle S(f)\rangle = \langle S(g)\rangle$, and there is a continuous map $F:cl[f(R)] \to cl[g(R)]$ with $g = F\circ f$, then F must be a homeomorphism.

Corollary 5.3: If X,Y are locally connected almost periodic minimal sets in R^n or C^n, then $\langle S(X)\rangle = \langle S(Y)\rangle$ if and only if X and Y are flow isomorphic.

Auslander and Hahn [1, page 136] have proven a result similar to Corollary 5.3. In their theorem X,Y are the orbit closures of almost periodic function in the space of bounded uniformly continuous complex-valued functions on R with the topological of uniform convergence.

BIBLIOGRAPHY

1. J. Auslander and F. Hahn, Point transitive flows, algebras of functions and the Bebutor system, Fund. Math. 60 (1967), 117-137.

2. A. S. Besicovitch, Almost Periodic Functions, Dover Pub., Inc., 1954.

3. M. L. Cartwright, Almost periodic flows and solutions of differential equations, Proc. London Math. Soc., 17 (1967), 355-380.

4. R. Ellis, Lectures on Topological Dynamics, W. A. Benjamin, Inc., New York, 1969.

5. W. H. Gottschalk and G. A. Hedlund, Topological Dynamics, Amer. Math. Soc. Coll. Pub. 36, Providence, R.I., 1955.

6. N. Minorsky, Nonlinear Oscillations, Van Nostrand Pub., Princeton, N. J., 1962.

7. J. Moser, On the theory of quasiperiodic motions, SIAM Review, 8 (1966), 145-172.

8. V. V. Nemytskii and V. V. Stepanof, Qualitative Theory of Differential Equations, Princeton University Press, Princeton, N. J., 1960.

9. L. S. Pontryagin, Topological Groups (2nd Edition), Gordon and Breach, Science Pub., New York, 1966.

Joseph F. Kent
University of Florida
Gainesville, Florida 32601

CHOQUET THEORY AND ERGODIC MEASURES FOR COMPACT GROUP EXTENSIONS

Harvey B. Keynes* and Dan Newton

§1. INTRODUCTION.

In this paper, we shall be concerned with the following problem: given a transformation group (X,T) which is a free group extension of another transformation group (Y,T), describe the structure of the set of invariant Borel probability measures of (X,T) which project onto an ergodic measure on (Y,T) in terms of the ergodic measures $\mathcal{E}(Y,T)$ of (Y,T), and the fibre group G. This problem can be further divided into two parts:

a) describe the geometric structure of this set in the sense of Choquet, and determine if one can obtain genuine ergodic decompositions, and

b) describe the "internal" structure of the subset $\mathcal{E}(X,T)$ of the ergodic measure on (X,T), i.e., determine a method for constructing an ergodic measure on (X,T) in terms of its ergodic projection on (Y,T) and the group G.

We shall prove a rather complete answer to a) under the assumption that X and G are compact Hausdorff, and that (X,T) admits an invariant Borel probability measure (this latter condition is satisfied by fairly general assumptions such as T is discretely amenable, or (X,T) is distal), and several observations concerning b), culminating again in

*This research is supported in part by NSF Grant GP-29321.

a rather complete answer under the additional assumptions of metrizability of X, G, and T, and commutativity of G. We emphasize that in general, very little is known about genuine ergodic decompositions on compact Hausdorff spaces, due to the fact that, in general, the ergodic measures do not form a Baire subset of the space of all invariant Borel probability measures. In addition, we shall make some observations concerning ergodic measures on inverse limit transformation groups. Combining this result with part a), we then show that if a transformation group (Z,T) is built from (W,T) via group extensions and inverse limits, then $\mathcal{E}(Z,T)$ is built from $\mathcal{E}(W,T)$ in the same way (i.e., there is a natural correspondence). This external description of $\mathcal{E}(Z,T)$ can be further amplified to yield a precise internal description under the assumptions of metrizability and commutativity.

§2. RESULTS.

Definition 1. *i)* The transformation group (X,T) is a *free G-extension* of (Y,T) if the map $G \times X \longrightarrow X$, $(g,x) \longrightarrow gx$ induces a jointly continuous left action G on X such that $g(xt) = (gx)t$ $(x \in X, g \in G, t \in T)$, $gx = x$ for some $x \in X$ implies $g = e$, and the orbit transformation group (X/G,T) is isomorphic to (Y,T). We write $\pi: (G;X,T) \longrightarrow (Y,T)$, where $\pi: X \longrightarrow Y$ is the canonical homomorphism.

ii) Denote by $\mathcal{P}(G)$ the collection of Borel probabilities on G, and let $\mathcal{M}(X,T)$ denote the set of invariant measures on (X,T). Let $\nu \in \mathcal{P}(G)$, $\mu \in \mathcal{M}(X,T)$. Then the *convolution of* ν

and μ, denoted $\nu*\mu$, is defined by

$$\nu*\mu(f) = \int_G \int_X f(gx)\, d\mu(x)\, d\nu(g)$$

($f \in \mathcal{C}(X)$). Here we use the notation $\omega(f) = \int_X f(x)\, d\omega(x)$, if $\omega \in \mathcal{M}(X,T)$, $f \in \mathcal{C}(X)$, the continuous functions on X.

<u>Lemma 2</u>. *i) The map* $*: \mathcal{P}(G) \times \mathcal{M}(X,T) \longrightarrow \mathcal{M}(X,T)$ *induces a jointly continuous action of the semigroup* $\mathcal{P}(G)$ *(with the operation of convolution in* $\mathcal{P}(G)$*) on* $\mathcal{M}(X,T)$.

ii) If $g \in G$, *and* δ_g *is point mass at* g, *we denote* $\delta_g*\mu$ *by* $g\mu$, *if* $\mu \in \mathcal{M}(X,T)$. *Then* $\mathcal{P}(G)*\mu$ *is a compact convex subset of* $\mathcal{M}(X,T)$ *with extreme points* $\{g\mu \mid g \in G\}$ *and containing exactly one* G*-invariant medsure, namely,* $\lambda*\mu$, *where* λ *is Haar measure on* G *(assuming* $g\mu = \mu$ *means* $g = e$*).*

We now note that the map $\pi: (X,T) \longrightarrow (Y,T)$ induces a map $\pi^*: \mathcal{M}(X,T) \longrightarrow \mathcal{M}(Y,T)$. Let $m \in \mathcal{E}(Y,T)$ and set $P_m = \{\mu \in \mathcal{M}(X,T) \mid \pi^*\mu = m\}$. Since (X,T) is a G-extension of (Y,T), it is well known that $P_m \neq \emptyset$, and that P_m is compact, convex.

The next result yields the geometry of P_m.

<u>Theorem 3</u>.

i) P_m *is a simplex.*

ii) P_m *is a* $\mathcal{P}(G)$*-invariant subset.*

iii) The set $E(P_m)$ *of extreme points of* P_m *satisfy the relationship* $E(P_m) = P_m \cap \mathcal{E}(X,T)$.

iv) If $\nu, \mu \in P_m \cap \mathcal{E}(X,T)$, *then there exists* $g \in G$ *with* $\nu = g\mu$. *Thus, if* $\nu \in P_m \cap \mathcal{E}(X,T)$, $E(P_m) = \{g\nu \mid g \in G\}$.

v) If $\nu \in E(P_m)$, *then* $P_m = \mathcal{P}(G)*\nu$.

Note that by *iv)* of Theorem 3, $E(P_m)$ is a closed subset of P_m. It follows from general Choquet theory that given $\omega \in P_m$, any representing measure for ω is concentrated on $E(P_m)$, i.e., the ergodic measures in P_m. Thus, we obtain genuine ergodic decompositions.

Corollary 4. *i) If* (Y,T) *is uniquely ergodic, then* $\mathfrak{M}(X,T)$ *is a Choquet simplex, i.e., has genuine ergodic decompositions.*

ii) P_m *is a singleton if and only if given* $\nu \in E(P_m)$, *the Haar convolution* $\lambda*\nu$ *is ergodic if and only if* $\tilde{m}$ *is ergodic, where* $\tilde{m} \in P_m$ *is defined by*

$$\tilde{m}(f) = \int_X \int_G f(gx)\, d\lambda(g)\, d\pi^{-1}m(g)$$

$(f \in \mathcal{C}(X))$, *and* $\pi^{-1}m$ *is the measure on the subalgebra* $\pi^{-1}\mathcal{B}(Y)$ *of the Borel sets* $\mathcal{B}(X)$ *of* X *given by* $(\pi^{-1}m)(\pi^{-1}B) = m(B)$.

This latter result is a generalization of a result of Parry [2].

Definition 5. Let H be a closed subgroup of G. Then $\pi = \pi_2 \cdot \pi_1$ where $\pi_1 : (H;X,T) \longrightarrow (X/H\,,\,T)$ is an H-group extension, and $\pi_2 : (X/H\,,\,T) \longrightarrow (Y,T)$ satisfies that each fibre is homeomorphic to the homogenous space G/H. Let $m \in \mathcal{E}(Y,T)$, and $\nu \in P_m$. Then ν is an *H-extension* of m if $\nu = \lambda_1 * (\pi_1^* \nu)$, where λ_1 is Haar measure on H, and $\pi_2 : (X/H, T, \pi_1^* \nu) \longrightarrow (Y, T, m)$ is a one-to-one homomorphism mod 0 (i.e., on sets of full measure).

Theorem 6. *Let* X be metric, G *be abelian and metric, and* T *locally compact separable. Let* $m \in \mathcal{E}(Y,T)$. *Then there exists a closed subgroup* H *of* G *such that if* $\nu \in P_m \cap \mathcal{E}(X,T)$, *then* ν *is an H-extension of* m.
In addition, π_2 *is actually an isomorphism* mod 0.

Corollary 7. *Let* G *be the circle and* X *metric. If* $\nu \in P_m \cap \mathcal{E}(X,T)$, *then* ν *is either isomorphic to* m *or is a finite-to-one extension of* m.

This result is originally due to Furstenberg [1].

With respect to inverse limits, we have the following result.

Theorem 8. *Let* $((X_\alpha, T) \mid \alpha \in I)$ *be an inverse system of transformation groups and* $(\mu_\alpha \mid \alpha \in I)$ *an inverse system of ergodic measures (i.e.,* $\mu_\alpha \in \mathcal{E}(X_\alpha, T)$ $(\alpha \in I)$ *and if* $\beta \geq \alpha$, $(\pi_\alpha^\beta)^* \mu_\beta = \mu_\alpha$*). Let* $(X,T) = \mathrm{invlim}(X_\alpha, T)$, *and* $\mu \in \mathcal{M}(X,T)$ *satisfying* $\pi_\alpha^* \mu = \mu_\alpha$ $(\alpha \in I)$. *Then* $\mu \in \mathcal{E}(X,T)$.

As a consequence, we have that given an inverse system $(\mu_\alpha \mid \alpha \in I)$ of ergodic measures, there is precisely one measure $\mu \in \mathcal{M}(X,T)$ such that $\pi_\alpha^* \mu = \mu_\alpha$ $(\alpha \in I)$, and it is ergodic.

Combining *iv*) of Theorem 3 and Theorem 8, we obtain:

Theorem 9. *Let* (X,T) *and* (Y,T) *be transformation groups such that there exists an ordinal* η *and a family* $((X_\gamma,T) \mid \gamma \leq \eta)$ *of transformation groups satisfying:*

1) $(X_0,T) = (Y,T)$

2) *If* $\gamma \leq \eta$ *and* γ *is not a limit ordinal, then* $(X_{\gamma+1},T)$ *is a* G_γ*- free group extension of* (X_γ,T).

3) *If* $\gamma \leq \eta$ *and* γ *is a limit ordinal, then* $(X_\gamma,T) = \operatorname{invlim}_{\beta<\gamma} (X_\beta,T)$.

4) $(X,T) = (X_\eta,T)$.

Then the family $(\mathcal{E}(X_\gamma,T) \mid \gamma \leq \eta)$ *satisfies:*

i) $\mathcal{E}(X_0,T) = \mathcal{E}(Y,T)$.

ii) *If* $\gamma \leq \eta$, *and* γ *is not a limit ordinal, then* $\mathcal{E}(X_{\gamma+1})$ *is a* G_γ - *group extension of* $\mathcal{E}(X_\gamma,T)$.

iii) *If* $\gamma \leq \eta$ *and* γ *is a limit ordinal, then* $\mathcal{E}(X_\gamma,T) = \operatorname{invlim}_{\beta<\gamma} \mathcal{E}(X_\beta,T)$.

iv) $\mathcal{E}(X_\eta,T) = \mathcal{E}(X,T)$.

In regard to the internal structure, we have by Theorem 6:

Theorem 10. *Let* (X,T) *and* (Y,T) *be as in Theorem* 9, *with* X *metric,* G_γ *abelian* $(\gamma \leq \eta)$ *and* T *locally compact separable. Let* (Y,T) *be uniquely ergodic, and* $\{m\} = \mathcal{E}(Y,T)$. *Let* $\mu \in \mathcal{E}(X,T)$. *Then there exist families* $(H_\gamma \mid \gamma \leq \eta,\ \gamma$

not a limit ordinal) *and* $(\mu_\gamma \mid \gamma \leq \eta)$ *such that* H_γ *is a closed subgroup of* G_γ $(\gamma \leq \eta,$ γ not a limit ordinal) *and* $\mu_\gamma \in \mathcal{E}(X_\gamma, T)$ $(\gamma \leq \eta)$ *satisfying:*

i) $\mu_0 = m$

ii) *If* $\gamma \leq \eta$ *and* γ *is not a limit ordinal, then* $\mu_{\gamma+1} = \lambda_\gamma * \mu_\gamma$, *where* λ_γ *is Haar measure on* H_γ.

iii) *If* $\gamma \leq \eta$ *and* γ *is a limit ordinal, then* $\mu_\gamma = \operatorname{invlim}_{\beta<\gamma} \mu_\beta$.

iv) $\mu_\eta = \mu$.

Remarks. 1) By modifying an example due to Furstenberg [1], we can obtain many of the different types of possible ergodic measures for circle extensions.

2) One can obtain a few results concerning lifting of entropy and maximal entropy in the case that T is the integers.

3) There are similar results for minimal sets in free simple (in the sense of topology; cf. [2]) G-extensions, under the assumption that G is abelian.

UNIVERSITY OF MINNESOTA

AND

UNIVERSITY OF SUSSEX

REFERENCES

1. H. Furstenberg, *Strict ergodicity and transformations of the torus*, Amer. J. Math. 83(1961), 573-601.

2. W. Parry, *Compact abelian group extensions of discrete dynamical systems*, Z. Wahr. verw. Geb. 13(1969), 95-113.

THE INTERMEDIATE TRANSFORMATION GROUPS

Ping-Fun Lam
Institute for Advanced Study

What will be seen here are some results which have been obtained, by us mostly, on a class of transformation groups which we here refer to as "intermediate", in a sense that we shall make clear. On the way we will also point out a few problems which associate themselves naturally with this study.

Consider a transformation group (X, T), where we assume that the phase space X is metric and the phase group T is arbitrary. Let d be the metric on X and let $N = N(X, T, d)$ be the subset of points of X where the (transition) group of homeomorphisms T fails to be equicontinuous with respect to d. The set N separates transformation groups with metrizable phase spaces in the following sense. Let $\mathfrak{M}$ be the class of all transformation groups each one of which has a metric phase space as well as a designate metric. By means of the set N we may divide $\mathfrak{M}$ into three subclasses.

Class 1. $N = \phi$.

Class 2. $N = X$.

Class 3. $X \neq N$, $N \neq \phi$.

The extreme classes 1 and 2 have provided some of the most interesting theories and examples in topological dynamics. To name a few we mention that a compact group acts on a metric space, or that a uniformly almost periodic transformation group built on a metric space are examples of class 1. Examples of class 2 are expansive transformation groups on self-dense metric spaces and distal flows on metric spaces which are free of equicontinuous points. When the set N is in general position, the corresponding transformation group is that of class 3 and is here termed intermediate.

Needless to say there have been an enormous number of works which have contributed directly or indirectly to the study of class 3. Perhaps, we will make our objective clear by first pointing out a few problems, which are among those we have in mind for the intermediate class.

Problem A. The structure of the set N.

Problem B. Given N (and T) determine the phase space X.

Problem C. For a given space X and a given topological group T determine N so that corresponding to the homeomorphs of N there are at most finite distinct isomorphic classes of actions of T on X--or other problems of this sort.

We now proceed to describe and illustrate by results for Problems A, B and C, in such an order.

Problem A. The structure of the set N includes, though is not limited by, its topological and dynamical properties. The following is a natural question to ask. If G is an arbitrary group of homeomorphisms on a metric space, what is its set of non-equicontinuous points like? A nearly complete answer for the case when N is totally disconnected is now to be given. As of most of the remaining results we assume X to be locally compact and connected. The following theorem was announced in [5] and proved in [7].

THEOREM 1. *Let* X *be locally compact and connected and let an intermediate transformation group* (X, T) *have the following properties:*

(a) N *is totally disconnected.*

(b) *If* $p \in N$ *then its orbit closure is compact.*

(c) The set $E = X-N$ has property (I) (to be defined in the following): Then N must be one of the following five kinds:

(i) a fixed point;

(ii) a union of two fixed points;

(iii) an orbit of two elements;

(iv) a minimal set which is homeomorphic to the Cantor set;

(v) a topological Cantor set which has a unique fixed point and all other orbits are dense in N.

The property (I) required for the set E in (c) of Theorem 1 is defined as follows.

Property (I) for E. If $\lim_{n\to\infty} y_0 t_n = p \in N$ for some $y_0 \in E$ and for a sequence $\{t_n\}$ in T, then $\lim_{n\to\infty} yt_n = p$ for every $y \in E$.

With the assumptions for Theorem 1 except (c), condition (c) holds under a number of topological conditions (cf. [6]). We mention two of these conditions.

α) E is a domain (i. e., open and connected).

β) E is connected and X is locally connected.

For a metric space, which is not necessarily locally compact and connected, it can be shown that the invariant set N, though not necessarily closed in general, is always a countable union of closed invariant sets [5], [7]. It follows that if X is also locally compact, then N is totally disconnected if and only if it is zero-dimensional. If X is a connected manifold of dimension at least 2, then (c) of Theorem 1 is automatically satisfied and if X is further assumed to be compact, then (b) in it is also satisfied. For closed n-manifolds $(n \geq 2)$ condition (a) is then the only hypothesis of Theorem 1.

The property (I), which we have referred to as a condition of indivisibility in [5], [6] and [7], is preferable to either one of conditions α) and β) above. It is preferable to condition α), since it allows Theorem 1 to include cases such as that when X is a closed n-manifold $(n \geq 2)$ and when N is not known to be closed. The reason that it is preferable to condition β) will be given after the next theorem is stated. There are simpler techniques for proving Theorem 1 for special cases such as when N is assumed consisting of only fixed points or when X is assumed to be locally connected. A proof for the general case is more difficult.

We turn to the case when N is not necessarily zero-dimensional. In [7] we identify the components of N to points to obtain a so-called partition or quotient transformation group (X_*, T) which, under suitable condition, satisfies the hypothesis of Theorem 1. In this way or by other similar means Theorem 1 is applicable to general cases. The method is not as simple as it may seem, for the canonical projection from X to X_* generally carries the set N of X to a superset of the set of non-equicontinuous points of X_*, rather than onto it. We state a result which is proved in a more general form in [7] by this technique.

The set N is said to be _dynamically disconnected_ if it is the union of two disjoint non-empty relatively closed invariant subsets. The definition is not to be confused with that for _dynamically decomposable_ set given in Nemytskii [8], where such a set is defined to be one which is representable as the union of two relatively closed invariant sets neither of which equals the set.

THEOREM 2. _Let_ (X, T) _be an intermediate transformation group, where_ X _is compact connected and where_ N _is closed and its complement_ E _is connected and has a point which is not almost periodic. If_ N _is dynamically_ disconnected, then N is the union of two of its invariant components.

Returning to the discussion of condition (c) for Theorem 1, we observe that even if the space X is assumed to be a manifold to begin with, the partition space X_* is in general not locally connected. For application on cases such as the proof of Theorem 2 the property (I) of Theorem 1 is then preferable as an hypothesis to condition β) given above.

In Husch and Lam [3] some interesting results on intermediate discrete flows are obtained, under the assumption that the set of nonwandering points is 0-dimensional. We state one of the principal results.

THEOREM 3. Let (X, T) be a discrete flow, where X is a closed n-manifold. If N is closed and non-empty and the set of nonwandering points is zero-dimensional, then the following properties hold.

(a) N is a continuum provided X is not homeomorphic to the n-sphere.

(b) N is either a continuum or a set consists of two fixed points when either X is homeomorphic to an even dimensional sphere or when X is homeomorphic to an odd dimensional sphere and (X, T) has a periodic point.

Although the proof of Theorem 3 does not separate cases according to the dimension of N, we remark that the harder cases are when N is of codimension 1 or 0.

Problem B. The following special theorem of [7] shows how the general objective of this problem may be achieved.

THEOREM 4. Let X be a connected n-manifold ($n \geq 2$). If the set N *is non-empty, totally disconnected and the orbit closure of its every point is compact, then X is topologically either an n-sphere, open n-cell, closed n-cell or the closed half of the Euclidean space* $\mathbb{R}^n$.

The case when N is not totally disconnected is presently under investigation.

Problem C. We also illustrate this problem by a theorem.

THEOREM 5. *Let h be an orientation-preserving homeomorphism of the Euclidean n-space* $\mathbb{R}^n$ *onto itself* $(n \neq 4, 5)$ *and let* h^* *be its unique extension to the n-sphere* S^n. *If the discrete flow generated by* h^* *is equicontinuous except at exactly two points, then h is topologically conjugate to the dilation* $(x \to 1/2\, x)$.

Theorem 5 is proved for $n = 2$ by Kerékjártó [4], for $n = 3$ by Homma-Kinoshita [1] and for $n > 5$ by Husch [2].

In closing we mention a few other interesting problems and questions for this study.

Problem D. Under the hypothesis of Theorem 1 the set N is closed. Find general conditions which guarantee the closedness of N. There are some results in this respect in [7].

Problem E. Class 3 is characterized by the simultaneous existence of both an equicontinuous point and a non-equicontinuous point. It is in general easy to give conditions that a non-equicontinuous point exists. What is the criterion for existence of an equicontinuous point, say for discrete flows? It can be shown that a discrete flow on a compact graph always has an equicontinuous point. What general spaces have this property?

In order to make the presentation simple we have omitted mentioning some interesting relevant works of other authors in this subject. We refer the interested readers to our introductions and references in [5], [6] and [7].

REFERENCES

[1] Homma, T., Kinoshita, S., "On a topological characterization of the dilation in E^3", Osaka Math. J. 6 (1954), 135-144.

[2] Husch, L. S., "A topological characterization of the dilation in E^n", Proc. Amer. Math. Soc. 28 (1971), 234-236.

[3] Husch, L. S., Lam, Ping-Fun, "Homeomorphisms of manifolds with zero-dimensional sets of nonwandering points", (to appear).

[4] Kerékjártó, B. v., "Topologische Charakterisierung der linearen Abbildungen", Acta Litt. Acad. Sci. Szeged, 6 (1934), 235-262.

[5] Lam, Ping-Fun, "On a theorem of B. von Kerékjártó", Bull. Amer. Math. Soc., 77 (1971), 230-234.

[6] ______________, "Equicontinuity and indivisibility in transformation groups", to appear in Trans. Amer. Math. Soc.

[7] ______________, "Almost equicontinuous transformation groups", (to appear).

[8] Nemytskii, V. V., "Topological problems of the theory of dynamical systems", Amer. Math. Soc. Transl., No. 103 (1954).

THE STRUCTURE OF COMPACT CONNECTED GROUPS WHICH ADMIT AN EXPANSIVE AUTOMORPHISM

WAYNE LAWTON

WESLEYAN UNIVERSITY

1. Remark We will make extensive use of the Pontryagin Duality Theory for compact abelian and discrete abelian groups. This material can be found in [17, Chapter 5] and in [12, Chapter 6]. We also assume a knowledge of torsion free rank for discrete abelian groups as described in [12, Appendix A]. When we use the term rank we mean torsion free rank. In particular, we use the following fact:

2. Theorem. Let G be a compact abelian group and let $\hat{G}$ be the character group of G. Then the dimension of G is equal to the rank of $\hat{G}$.

Proof. [12, Theorem 24.28, page 386] and [17, Example 49, page 148].

3. Remark. We assume a knowledge of topological entropy, a notion introduced in [2] as an invariant for a homeomorphism of a compact topological space onto itself. Let X be a compact topological space and let φ be a homeomorphism of X onto itself. Then the topological entropy of φ, which will be denoted $H(\varphi)$, satisfies the following basic properties:

4. Theorem. Let φ be the shift on X^Z where X is an infinite compact space. Then $H(\varphi) = \infty$.

Proof. [2, pages 315-316].

5. Theorem. Let X be a compact space and let φ be an expansive homeomorphism of X onto X. Then $H(\varphi) < \infty$.

Proof. [14, Theorems 2.6 and 3.2]

6. Theorem. Let X and Y be compact spaces, let φ be a homeomorphism of X onto X, and let θ be a homeomorphism of Y onto Y. Let $\Psi : X \longrightarrow Y$ be a continuous onto map such that $\Psi\varphi = \theta\Psi$. Then $H(\varphi) \geq H(\theta)$.

Proof. This follows from the fact that an open cover of Y can be lifted back to an open cover of X by Ψ^{-1} and the way in which entropy is defined in [2] in terms of open covers.

7. Remark. The notion of topological entropy arose from a measure theoretic analogue for measure-preserving transformations of a finite measure space. It was conjectured in [2, page 318] that the two notions of entropy coincide for automorphisms of a compact separable group. Under the assumption that the group in question is abelian, this conjecture was proved true in [3].

8. Standing Notation. For the remainder of this chapter, K will denote the compact circle group. Let G be a compact abelian group, let A be an automorphism of G, and let H be an A-invariant subgroup of G. The character group of G will be denoted by $\hat{G}$, the canonical automorphism of $\hat{G}$ induced by A will be denoted by $\hat{A}$, and the automorphism of H obtained by restricting A to H will be denoted by A_H. The identity element of any group will be denoted by 0 and all groups will be considered to be additive groups. If X and Y are sets with $X \subset Y$, we will denote the complement of X in Y by $Y \longrightarrow\!\!| \; X$. R denotes the set of all real numbers and Z denotes the set of all integers.

9. Definition. Let G be a compact abelian group and let A be an automorphism of G. We say that $\hat{G}$ is finitely generated with respect to $\hat{A}$ if there is a finite number of elements $x_1, \dots, x_n$ of $\hat{G}$ such that the union of the orbits of $x_1, \dots, x_n$ under $\hat{A}$ generates $\hat{G}$.

10. <u>Remark</u>. The condition above is equivalent to the condition that the union of the orbits of $x_1, \ldots, x_n$ under $\hat{A}$ separates the points of G, i.e. for every $g \in G$, $g \neq 0$ implies there exists x_i $(1 \leq i \leq n)$ and there exists $k \in Z$ such that $(\hat{A}^k x_i)g = x_i(A^k g) \neq 0$. This follows from [12, Theorem 23.20, pages 364].

11. <u>Lemma</u>. Let G be a compact abelian group and let A be an expansive automorphism of G. Then $\hat{G}$ is finitely generated with respect to $\hat{A}$.

<u>Proof</u>. Let V be an open expansive neighborhood of e in G, and let $G \dashv V$ denote the complement of V in G. Since characters separate points in G, for each $g \in G \dashv V$ there exists $x_g \in \hat{G}$ such that $x_g(g) \neq 0$. For each $g \in G \dashv V$ let U_g be an open neighborhood of g in G such that $h \in U_g$ implies $x_g(h) \neq 0$. Since $G \dashv V$ is compact we may choose a finite number of elements $g_1, \ldots, g_m \in G \dashv V$ such that $\{U_{g_i} \mid 1 \leq i \leq m\}$ covers $G \dashv V$. We need to show that the union of the orbits of $x_{g_1}, x_{g_2}, \ldots, x_{g_m}$ under $\hat{A}$ separates the points of G. Let $g \in G$ such that $g \neq 0$. Since V is an expansive neighborhood of e in G, there exists $k \in Z$ such that $A^k g \in G \dashv V$. Since $\{U_{g_i} \mid 1 \leq i \leq m\}$ covers $G \dashv V$, there exists i $(1 \leq i \leq m)$ such that $A^k g \in U_{g_i}$. Then $x_{g_i}(A^k g) = (A^k x_{g_i})(g) \neq 0$, and the lemma follows by 10.

12. <u>Definition</u>. Let G be a compact abelian group, and let A be an automorphism of G. We say that $\hat{G}$ <u>has locally finite rank with respect to</u> $\hat{A}$ if for every $x \in \hat{G}$, the subgroup of $\hat{G}$ generated by the orbit of x under $\hat{A}$ has finite rank. We note this condition on individual characters implies the condition on finite sets of characters.

13. Theorem. Let G be a compact connected abelian group and let A be an automorphism of G such that $H(A) < \infty$. Then $\hat{G}$ has locally finite rank with respect to $\hat{A}$.

Proof. Let $x \in \hat{G}$ and let $O(x) = \{\hat{A}^k x \mid k \in Z\}$ denote the orbit of x under $\hat{A}$. It suffices to show the existence of a finite subset $S \subset O(x)$ such that every element in $O(x)$ is integrally dependent on elements in S. We consider the following cases:

Case 1. The set $O(x)$ is integrally dependent, hence, there exists a positive integer k and there exist integers $a_1, \ldots, a_k$, bot all equal to zero, and there exist elements $x_1, \ldots, x_k \in O(x)$ such that $\Sigma_{i=1}^k a_i x_i = 0 \in \hat{G}$.

Case 2. The set $O(x)$ is not integrally dependent.

We will show that case 1 gives the conclusion and we will show that case 2 contradicts the hypothesis $H(A) < \infty$.

Assume case 1 and assume, without loss of generality, that $a_i \neq 0$ for all $1 \leq i \leq k$. Let $q_1, \ldots, q_k$ be integers such that $x_i = \hat{A}^{q_i}(x)$ for all $1 \leq i \leq k$. Let $L = \min\{q_i\}$ and let $M = \max\{q_i\}$. Let $S = \{\hat{A}^h(x) \mid L \leq h \leq M\}$ and let $\langle S \rangle$ denote the subgroup of $\hat{G}$ generated by S. It will suffice to show that for every element $y \in O(x)$, there exists an integer $c \neq 0$ such that $cy \in \langle S \rangle$. The above considerations ensure there exists $\{b_h \mid L \leq h \leq M\} \subset Z$ with $b_L \neq 0$ and $b_M \neq 0$ such that $\Sigma_{h=L}^M b_h \hat{A}^h(x) = 0$. Hence, applying $\hat{A}$ and $\hat{A}^{-1}$ to both sides gives $\Sigma_{h=L}^M b_h \hat{A}^{h+1}(x) = \hat{A}(0) = 0$ and $\Sigma_{h=L}^M b_h \hat{A}^{h-1}(x) = \hat{A}^{-1}(0) = 0$. Hence, $b_M \hat{A}^{M+1}(x) \in \langle S \rangle$ and $b_L \hat{A}^{L-1}(x) \in \langle S \rangle$. Hence, $b_M^2 \hat{A}^{M+2}(x) = b_M(b_M \hat{A}^{M+2}(x)) = b_M(-\Sigma_{h=L}^{M-1} b_h \hat{A}^{h+2}(x)) = b_M(-b_{M-1}\hat{A}^{M+1}(x)) + b_M(-\Sigma_{h=L}^{M-2} b_h \hat{A}^{h+2}(x)) \in \langle S \rangle$ and $b_L^2 \hat{A}^{L-2}(x) \in \langle S \rangle$. It follows by induction that for every integer $p > 0$, $b_M^p \hat{A}^{M+p}(x) \in \langle S \rangle$ and $b_L^p \hat{A}^{L-p}(x) \in \langle S \rangle$. Since every element of the orbit of x under $\hat{A}$ is integrally dependeny on the finite set S, it follows that every element of the subgroup of $\hat{G}$ generated by the orbit of x under $\hat{A}$ is integrally dependent on elements in the set S.

Assume case 2. Let $V = \{v_i \mid i \in Z\}$ be any set such that $v_i \neq v_j$ for $i \neq j$, and let F be the discrete free abelian group on V. Let σ be the shift automorphism on K^Z. It is well known that $K^Z \approx F$ and σ can be defined as the unique automorphism of F such that $\hat{\sigma}(v_i) = v_{i+1}$ for all $i \in Z$. Define a homomorphism $\Psi : F \longrightarrow \hat{G}$ as follows: let $\Psi(v_i) = \hat{A}^i(x)$ for all $i \in Z$ and extend uniquely to a homomorphism from F to $\hat{G}$. Let $y = \Sigma a_i v_i \in \mathrm{Ker}(\Psi)$ where a_i are integers such that $a_i \neq 0$ for finite number of i's. Then $\Psi(y) = \Sigma a_i \hat{A}^i(x) = 0$ implies $a_i = 0$ for all $i \in Z$ since there does not exist a nontrivial integral dependence in $O(x)$. Hence, $y = 0 \in F$ and we conclude Ψ is an isomorphism of F into $\hat{G}$. By the way Ψ was constructed, diagram one is commutative. If we

$$\begin{array}{ccc} \hat{G} & \xrightarrow{\hat{A}} & \hat{G} \\ {\scriptstyle\Psi}\uparrow & & \uparrow{\scriptstyle\Psi} \\ F & \xrightarrow{\hat{\sigma}} & F \end{array}$$

diagram one.

$$\begin{array}{ccc} G & \xleftarrow{A} & G \\ {\scriptstyle\hat{\Psi}}\downarrow & & \downarrow{\scriptstyle\hat{\Psi}} \\ K^Z & \xleftarrow{\sigma} & K^Z \end{array}$$

diagram two.

dualize diagram one, we get diagram two, which is also commutative. Now since Ψ is an isomorphism into, $\hat{\Psi}$ is a homomorphism onto. By 6 and 4, $H(A) \geq H(\sigma) = \infty$. This contradiction completes the proof.

14. <u>Theorem</u>. Let G be any compact group and let there exist an expansive automorphism A of G. Then G has finite dimension.

<u>Proof</u>. Since the connected component of the identity element in G is a compact A-invariant subgroup of G with the same dimension as G, [16, pages 236-239], we can assume without loss of generality that G is connected. Hence, the result in

[15, Theorem 3.2, page 135] implies G is abelian. By 11, $\hat{G}$ is finitely generated with respect to $\hat{A}$. By 5, $H(A) < \infty$, and hence by 13, $\hat{G}$ has locally finite rank with respect to $\hat{A}$. Hence, $\hat{G}$ has finite rank and thus by 2 the dimension of G is finite.

15. Definition. Let n be a positive integer, let F be an $n \times n$ matrix with integer entries such that the determinant of F is not zero, and let d be a positive integer. By the generalized solenoidal group of type (n, F, d) we mean the group S constructed as follows:

Consider elements of the n-dimensional torus $K^n = R^n/Z^n$ as being represented by n-tuple columns of real numbers. Let σ denote the shift automorphism of $(K^n)^Z = \{(y_i)_{i \in Z} | \forall i \in Z, y_i \in K^n\}$. Let G be the subset of $(K^n)^Z$ defined as follows:
$G = \{(y_i)_{i \in Z} \in (K^n)^Z \mid \forall i \in Z, dy_{i+1} = Fy_i\}$. We note that G is a closed σ-invariant subgroup of $(K^n)^Z$. The generalized solenoidal group S of type (n, F, d) will be defined to be the connected component of the identity of the group G. The shift automorphism of the generalized solenoidal group S will be the restriction to S of the shift automorphism σ of $(K^n)^Z$.

16. Lemma. Let n be a positive integer and let G be a compact connected abelian group with dimension n. Let A be an automorphism of G such that $\hat{G}$ is finitely generated with respect to $\hat{A}$. Then there exists a finite subset $\{x_1, \ldots, x_n\}$ of $\hat{G}$ with $x_i \neq x_j$ for $i \neq j$, satisfying the following conditions:

(1) The union of the orbits of $x_1, \ldots, x_n$ under $\hat{A}$ generates $\hat{G}$.

(2) The subgroup of $\hat{G}$ generated by the set $\{x_1, \ldots, x_n\}$ has rank n, hence, the set $\{x_1, \ldots, x_n\}$ is integrally independent.

Proof. Since $\hat{G}$ is finitely generated with respect to $\hat{A}$ there exists a positive integer q and elements $y_1, \ldots, y_q$ of $\hat{G}$

such that the union of orbits of $y_1, \ldots, y_q$ under $\hat{A}$ generates $\hat{G}$. Since G has dimension n, then by 2 $\hat{G}$ has rank n; hence there exist elements $y_{q+1}, \ldots, y_{q+n}$ of $\hat{G}$ such that the set $\{y_{q+1}, \ldots, y_{q+n}\}$ generates a subgroup of $\hat{G}$ with rank n. The union of the orbits of $y_1, \ldots, y_q, y_{q+1}, \ldots, y_{q+n}$ under $\hat{A}$ generates $\hat{G}$. Now we make use of the fact G is connected. The characters $y_1, \ldots, y_q, y_{q+1}, \ldots, y_{q+n}$ define in a natural way a homomorphism Ψ of G into the $(n+q)$-dimensional torus K^{q+n}. Since the subgroup of $\hat{G}$ generated by $\{y_1, \ldots, y_{q+n}\}$ has rank n, $\Psi(G)$ has dimension n.

Since G is compact and connected, $\Psi(G)$ is a compact connected subgroup of K^{q+n}. Therefore, $\Psi(G)$ is isomorphic to an n-dimensional torus K^n.

Let $\pi_1, \ldots, \pi_n$ be coordinate projections from $\Psi(G)$ to K and let $x_i = \pi_i \Psi : G \longrightarrow K$ for $1 \leq i \leq n$. The set $\{x_1, \ldots, x_n\}$ is easily seen to satisfy conditions 1 and 2. The proof is completed.

17. Lemma. Let n be a positive integer and let G be a compact connected abelian group with dimension n. Let A be an automorphism of G and let $\{x_1, \ldots, x_n\}$ be a finite subset $\{x_1, \ldots, x_n\}$ of $\hat{G}$, with $x_i \neq x_j$ for $i \neq j$, such that the set $\{x_1, \ldots, x_n\}$ generates a subgroup of $\hat{G}$ with rank n. For all $i \in Z$, let y_i denote the n-tuple column $\begin{pmatrix} \hat{A}^i x_1 \\ \vdots \\ \hat{A}^i x_n \end{pmatrix}$. Then there exists a positive integer d and n n matrix F with integer entries and non zero determinant, such that for all $i \in Z$, $dy_{i+1} = F\, y_i$.

Proof. Since rank $\hat{G} = n =$ rank of group generated by $\{x_1, \ldots, x_n\}$, for all $1 \leq j \leq n$, the set $\{x_1, \ldots, x_n, \hat{A}\, x_j\}$ is integrally dependent. Hence, for all $1 \leq j \leq n$, there exist integers $f_{j1}, \dots, f_{jn}, d_j$, not all equal to 0, such that $d_j \hat{A} x_j = \Sigma_{k=1}^{n} x_k$. Since the set $\{x_1, \dots, x_n\}$ is integrally independent, each $d_j \neq 0$, hence without loss of generality we can assume there exists a positive integer d such that $d_j = d$ for all $1 \leq j \leq n$. Let F be the $n \times n$ matrix (f_{jk}), $1 \leq j, k \leq n$. Then clearly $d\, y_1 = d\, \hat{A}\, y_o = F\, y_o$. Since for all $i \in Z$,

$$y_i = \begin{pmatrix} \hat{A}^i x_1 \\ \cdot \\ \cdot \\ \cdot \\ \hat{A}^i x_n \end{pmatrix} = \hat{A}^i \begin{pmatrix} x_1 \\ \cdot \\ \cdot \\ \cdot \\ x_n \end{pmatrix} = \hat{A}^i y_o$$

, it follows that for all $i \in Z$, $dy_{i+1} = d\, \hat{A}^i \hat{A} y_o = \hat{A}^i dy_1 = \hat{A}^i Fy_o = Fy_i$. We need to show that the determinant of F is non zero. By repeating the similar argument there exists a positive integer c and an $n \times n$ matrix E with integer entries, such that for all $i \in Z$, $c\, y_i = E\, y_{i+1}$. Then in particular, $d\, c\, y_o = dEy_1 = E\, dy_1 = EFy_o$. Since

$$y_o = \begin{pmatrix} x_1 \\ \cdot \\ \cdot \\ x_n \end{pmatrix}$$

and the set $\{x_1, \dots, x_n\}$ is integrally independent, it follows that F is an invertible martix with inverse $\frac{1}{dc} E$. Hence, the determinant of F is non zero and the proof is completed. We note that the inverse of F is in general an $n \times n$ rational valued matrix, i.e. the determinant of F may have absolute value greater than 1.

18. Lemma. Let n be a positive integer. Let $V = \{v_i^k \mid i \in Z, 1 \leq k \leq n\}$ where $v_{i'}^{k'} \neq v_i^k$ for $k' \neq k$ or $i' \neq i$. Let J be the free abelian group generated by V. Let ξ denote the automorphism of J defined by:

$$\xi(v_i^k) = v_{i+1}^k \quad \text{for all } i \in Z,\ 1 \leq k \leq n.$$

Let I_1, I_2 be subgroups of J satisfying the following conditions:

1) I_1 and I_2 are ξ-invariant.

2) The quotient group J/I_1 is torsion free.

3) $I_1 \subset I_2$.

4) There exists a positive integer d and an $n \times n$ matrix F with integer entries and non zero determinant, such that for all $i \in Z$, I_1 contains each element of the following n-tuple column: $d \begin{pmatrix} v^1_{i+1} \\ \vdots \\ v^n_{i+1} \end{pmatrix} - F \begin{pmatrix} v^1_i \\ \vdots \\ v^n_i \end{pmatrix}$.

5) If $a_1, \ldots, a_n$ are integers such that $\Sigma^n_{k=1} a_k v^k_o \in I_2$, then $a_k = 0$ for all $1 \leq k \leq n$. Then $I_1 = I_2$.

Proof. Let $\varphi_1 : J \longrightarrow J/I_1$ and $\varphi_2 : J \longrightarrow J/I_2$ denote the canonical epimorphisms. By conditions 1 and 4, for every element $u \in J$, there exists a non-zero integer $q \in Z$, and there exists an element v in the subgroup of J generated by $\{v^1_o, v^2_o, \ldots, v^n_o\}$, such that $qu - v \in I_1$. We refer the reader to the proof of case 1 of 13, since the argument is similar. Now let $u \in I_2$ and let q and v be chosen as above. Then $qu \in I_2$, and condition 3 implies $v \in I_2$. Hence by condition 5, $v = 0$. Then $qu \in I_1$ implies $q\,\varphi_1(u) = \varphi_1(qu) = 0 \in J/I_1$, hence by condition 2, $\varphi_1(u) = 0 \in J/I_1$. Thus $u \in I_1$ and the proof is completed.

19. Theorem. Let G be a compact connected abelian group. Let A be an automorphism of G such that $\hat{G}$ is finitely generated with respect to $\hat{A}$, and such that $\hat{G}$ has locally finite rank with respect to $\hat{A}$. Then there exist positive integers n and d, and there exists an $n \times n$ matrix F with integer entries and non zero determinant such that G is isomorphic to the generalized solenoidal group of type (n,F,d) and A corresponds to the shift automorphism σ.

Proof. Since $\hat{G}$ has finite rank, by 2, G has finite dimension. Let $n = \dim(G)$. By 16, there exists a finite subset $\{x_1, \dots, x_n\}$ of $\hat{G}$, with $x_i \neq x_j$ for $i \neq j$, satisfying the following conditions:

1) The union of the orbits of $x_1, \dots, x_n$ under $\hat{A}$ generates $\hat{G}$.

2) The set $\{x_1, \dots, x_n\}$ generates a subgroup of $\hat{G}$ of rank n.

Now 17 and condition 1 imply there exists a positive integer d and an $n \times n$ matrix F with integer entries and non-zero determinant, such that for all $i \in Z$,

$$d \begin{pmatrix} \hat{A}^{i+1}x_1 \\ \vdots \\ \hat{A}^{i+1}x_n \end{pmatrix} = F \begin{pmatrix} \hat{A}^{i}x_1 \\ \vdots \\ \hat{A}^{i}x_n \end{pmatrix} .$$

Define a homomorphism $\Psi : G \longrightarrow (K^n)^Z$ as follows: for all $g \in G$, let $\Psi(g) = (y_i)_{i \in Z} \in (K^n)^Z$ where for all $i \in Z$,

$$y_i = \begin{pmatrix} (\hat{A}^i x_1)g \\ \vdots \\ (\hat{A}^i x_n)g \end{pmatrix} \in K^n.$$

Let $S \subset (K^n)^Z$ denote the generalized solenoidal group of type (n,F,d). Clearly $G \subset S$ and if σ denotes the shift on S, then $\Psi A = \sigma \Psi : G \longrightarrow S$. We need only show that $G = S$. Let $V = \{v_i^k \mid i \in Z, 1 \leq k \leq n\}$ where $v_{i'}^{k'} \neq v_i^k$ for $k' \neq k$ or $i' \neq i$. Let J be the free group generated by V. Let ξ denote the automorphism of J defined by $\xi(v_i^k) = v_{i+1}^k$ for all $i \in Z$, $1 \leq k \leq n$. Define a map $\varphi_1 : J \longrightarrow \hat{S}$ as follows: for all $i \in Z$, $1 \leq k \leq n$, let $\varphi_1(v_i^k) = \pi_i^k|_S$ where $\pi_i^k : (K^n)^Z \longrightarrow K$ is defined as follows: for all $(y_i)_{i \in Z} \in (K^n)^Z$, $\pi_i^k[(y_i)_{i \in Z}] = y_i^k$ where

$$y_i = \begin{pmatrix} y_i^1 \\ \vdots \\ y_i^k \\ \vdots \\ y_i^n \end{pmatrix} \in K^n.$$

Clearly $\varphi_1 \xi = \hat{\sigma} \varphi_1 : J \longrightarrow \hat{S}$ and φ_1

is onto. Since condition 1 implies $\Psi : G \longrightarrow S$ is one-to-one, $\hat{\Psi} : \hat{S} \longrightarrow \hat{G}$ is onto and $\hat{\Psi}\hat{\sigma} = \hat{A}\hat{\Psi} : \hat{S} \longrightarrow \hat{G}$. Let $\varphi_2 = \hat{\Psi}\varphi_1 : J \longrightarrow \hat{G}$. Clearly, for all $i \in Z$, $1 \leq k \leq n$, $\varphi_2(v_i^k) = \hat{A}^i x_k$. Since $\varphi_2 \xi = \hat{A}\varphi_2 : J \longrightarrow \hat{G}$, if we let I_1 denote the kernel of $\varphi_1 : J \longrightarrow \hat{S}$ and if we let I_2 denote the kernel of $\varphi_2 : J \longrightarrow \hat{G}$, then I_1 and I_2 are ξ invariant. Clearly $I_1 \subset I_2$ and it will suffice to show $I_1 = I_2$. Since S is connected, by [12, 24.25, page 385], $J/I_1 \approx \hat{S}$ is torsion free. Condition 4 of 18 is valid for I_1 from the definition of I_1 and S. Finally, since the set $\{x_1, \ldots, x_n\}$ has rank n, and $\varphi_2(v_0^k) = x_k$ for $1 \leq k \leq n$, condition 5 of 18 is valid for I_2. Hence, by 18, $I_1 = I_2$. The proof is completed.

20. <u>Remark</u>. Clearly, if S is a generalized solenoidal group and σ is the shift automorphism on S, then $\hat{S}$ is finitely generated with respect to $\hat{\sigma}$ and $\hat{S}$ has locally finite rank with respect to $\hat{\sigma}$. By the argument in 14 and 19, if G is a compact connected group and if A is an expansive automorphism of G, then G is isomorphic to a generalized solenoidal group and A corresponds to the shift automorphism on the generalized solenoidal group. The next result characterizes those generalized solenoidal groups such that the shift automorphism is expansive.

21. <u>Theorem</u>. Let S be the generalized solenoidal group of type (n,F,d) and let σ be the shift automorphism of S. Then σ is expansive iff F has no eigenvalues of absolute value d.

<u>Proof</u>. Consider $\frac{1}{d}F$ as an automorphism of the real vector space R^n. By [7, Theorem 1.], $\frac{1}{d}F$ acts expansively on R^n if and only if $\frac{1}{d}F$ has no eigenvalues of absolute value 1, that is, F has no eigenvalues of absolute value d. Hence it is enough to show that σ is expansive on S if and only if $\frac{1}{d}F$ is expansive on R^n.

Let $G = \{(y_i)_{i \in Z} \in (K^n)^Z \mid \forall\, i \in Z,\ dy_{i+1} = F\, y_i\}$. Clearly, for all $(y_i)_{i \in Z} \in G$ and for all $i \in Z$, $cy_i = E\, y_{i+1}$. Let $\pi: R^n \longrightarrow K^n$ be the canonical homomorphism. Since π is a local isomorphism, there exist a local cross section φ of π which maps a neighborhood U of the identity $0 \in K^n$ isomorphically onto a neighborhood V of the identity $0 \in R^n$, $\varphi: U \longrightarrow V$, and $\pi\varphi: U \longrightarrow U$ is the identity map. Since the generalized solenoidal group S of type (n,F,d) is defined in 15 to be the connected component of the identity in G; we observe that we can choose U sufficiently small such that if $(y_i)_{i \in Z} \in G$, and if for all $i \in Z$, $y_i \in U$, then $(y_i)_{i \in Z} \in S$.

Assume $\frac{1}{d}F$ is not expansive on R^n. To show σ is not expansive on S, it will suffice to show that if W is any neighborhood of $0 \in K^n$, then there exists $(y_i)_{i \in Z} \in G$ such that for all $i \in Z$, $y_i \in W$, and such that $(y_i)_{i \in Z} \neq 0 \in G$. Since $\frac{1}{d}F$ is not expansive on R^n, there exist $\begin{pmatrix} x_1 \\ \vdots \\ x_n \end{pmatrix} \in R^n$ such that $\begin{pmatrix} x_1 \\ \vdots \\ x_n \end{pmatrix} \neq 0 \in R^n$ and such that for all $i \in Z$, $(\frac{1}{d}F)^i \begin{pmatrix} x_1 \\ \vdots \\ x_n \end{pmatrix} \in \varphi(W \cap U) \subset V$. For all $i \in Z$, let $w_i = (\frac{1}{d}F)^i \begin{pmatrix} x_1 \\ \vdots \\ x_n \end{pmatrix} \in \varphi(W \cap U)$ and let $y_i = \pi(w_i)$. Then $(y_i)_{i \in Z} \in K^n$ is our desired element.

Now assume σ is not expansive on S. Let V' be any neighborhood of $0 \in R^n$. Choose $(y_i)_{i \in Z} \in S$ such that $(y_i)_{i \in Z} \neq 0 \in S$ and such that for all $i \in Z$, $y_i \in \pi'(V' \cap V) \subset U$. For every $i \in Z$, let $w_i = \varphi(y_i) \in V' \cap V \subset R^n$. Observe that for every $i \in Z$, $(\frac{1}{d}F)w_i = w_{i+1}$. Hence $w_0 \in R^n$ is such that $w_0 \neq 0$

and for all $i \in Z$, $(\frac{1}{d}F)^i w_o = w_i \in V'$. Since V' was arbitrary, $\frac{1}{d}F$ is not expansive. The proof is completed.

REFERENCES

1. Abramov, L. M., The entropy of an automorphism of a solenoidal group, Theory of Probability and its Applications, vol. 4 (1959), pp. 231-236 (English translation).
2. Adler, R. M.; Konheim, A. G.; McAndrew M. H., Topological entropy, Transactions of the American Mathematical Society, vol. 114 (1965), pp. 309-319.
3. Aoki, N., Topological entropy and measure for automorphisms on compact groups, Mathematical Systems Theory, vol. 5, (1971), pp. 4-7.
4. Arov, D. Z., Topological similitude of automorphisms and translations of compact commutative groups, Uspehi Mathematiceskih Nauk, vol. 18 (1963), no. 5 pp. 133-138 (Russian).
5. Artin, M.; Mazur, B., On periodic points, Annals of Mathematics, vol. 81 (1965), pp. 82-99.
6. Auslander, J.; Gottschalk, W. H., (editors) Topological dynamics, An international symposium, Benjamin, New York, 1968.
7. Eisenberg, M., Expansive automorphisms of finite dimensional vector spaces, Fundamenta Mathematicae, vol. 59 (1966), pp. 307-312.
8. Eisenberg, M., Expansive transformation semi groups of endomorphisms, Fundamenta Mathematicae, vol. 59 (1966), pp. 313-321.
9. Gottschalk, W. H.; Hedlund, G. A., Topological dynamics, American Mathematical Society Colloquium Publications, vol. 36. American Mathematical Society, Providence, 1955.
10. Hedlund, G. A., Endomorphisms and automorphisms of shift dynamical systems, Mathematical Systems Theory, vol. 3 (1969), pp. 320-375.
11. Helgason, S., Differential geometry and symmetric spaces, Academic Press, New York, 1962.
12. Hewitt, E.; Ross K. A., Abstract harmonic analysis, vol. 1. Springer Verlag, Berlin; 1963.

13. Hofmann, K. H.; Mostert, P., Splitting in topological groups, Memoirs of the American Mathematical Society, no. 43. American Mathematical Society, Providence, 1963.

14. Keynes, H. B.; Robertson, J. B., Generators for topological entropy and expansiveness, Mathematical Systems Theory, vol, 3 (1969), pp. 51-59.

15. Lam, Ping-Fun, On expansive transformation groups, Transactions of the American Mathematical Society, vol. 150 (1970), pp. 131-138.

16. Montgomery, Deane; Zippin, Leo, Topological transformation groups, Interscience Tracts in Pure and Applied Mathematics, Tract 1, Interscience, New York, 1955.

17. Pontryagin, L., Topological groups, Princeton Mathematical Series, vol. 2. Princeton University Press, Princeton, 1939.

18. Reddy, William, The existence of expansive homeomorphisms on manifolds, Duke Mathematical Journal, vol. 32 (1965), pp. 494-509.

19. Smale, Stephen, Differentiable dynamical systems, Bulletin of the American Mathematical Society, vol. 73 (1967) pp. 747-817.

20. Walters, Peter, Introductory lectures on ergodic theory, lecture notes, University of Maryland, College Park, Maryland, 1970.

CHARACTERISTIC SEQUENCES

Nelson G. Markley

Let $\Omega = \{0,1\}^Z$ when Z denotes the integers and let σ denote the shift homeomorphism on the sequence space Ω. We will denote the nth coordinate of $x \in \Omega$ by $x(n)$, and then we have $\sigma(x)(n) = x(n+1)$. A sequence x in Ω will be called a characteristic sequence if x is an almost automorphic point of the cascade (Ω,σ). Our goal is to initiate a geometric study of characteristic sequences.

Let x_0 be a characteristic sequence and let $X = \mathrm{Cl}(\mathcal{O}(x_0))$ where as usual $\mathcal{O}(x) = \{\sigma^n x\colon n \in Z\}$. We will assume x_0 is not periodic. Then (X,σ) is an almost automorphic minimal set and its proximal relation equals its equicontinuous structure relation. There exists a monothetic topological group G, a generator g of G and a homomorphism p of (X,σ) onto (G, M_g) such that $p(x) = p(y)$ if and only if x and y are proximal where $M_g(z) = z + g$. Let $A = p(\{x \in X\colon x(0) = 1\})$. Then A has the following properties: $A = \mathrm{Cl}(\mathrm{Int}\, A)$, $\partial A \neq \emptyset$, $A + z = A$ implies $z = 0$. Let $\mathcal{C}(G)$ denote the subsets of G with these properties. Moreover, we have

$$x_0(n) = \chi_A(p(x) + ng)$$

which explains our choice of the name characteristic sequence. There is a converse to this. For $A \in \mathcal{C}(G)$ let $z_0 \in G \setminus \bigcup_{n=-\infty}^{\infty} \partial A + ng$, $x_0(n) = \chi_A(z_0 + ng)$, and $X_A = \mathrm{Cl}(\mathcal{O}(x_0))$.

Then (X_A,σ) determines A as above. We can now state the general problem. What information does the geometry of A contain about the dynamics of (X_A,σ)? The remainder of our discussion will consist of stating a few theorems of this type.

It is easy to see that given $A,B \in \mathcal{C}(G)$, $X_A = X_B$ if and only if $A = B + z$ for some z. The question of isomorphism is more delicate but can be answered in G which adds some substance to our general problem. Let A and B be elements of $\mathcal{C}(G)$ and let g be a generator of G. We say B can be g-constructed from A provided that

$$B = \bigcup_{i=1}^{m} U_i$$

such that $\operatorname{Int} U_i \cap U_j = \emptyset$ when $i \neq j$ and each U_i has the form

$$U_i = \left[\bigcap_{j=1}^{P(i)} (A+p_{ij}g)\right] \cap \left[\bigcap_{k=1}^{Q(i)} (A_0+q_{ik}g)\right]$$

where $A_0 = \mathrm{Cl}(G\setminus A)$ and the integers $p_{i1},\cdots,p_{i\,P(i)},q_{i1},\cdots,q_{i\,Q(i)}$ are all distinct. We say A and B are g-equivalent if A can be g-constructed from B and B can be g-constructed from A.

Theorem 1. Let g be a generator of G, let A and B be elements of $\mathcal{C}(G)$, and let X_A and X_B be the almost automorphic minimal sets obtained from A and B using g. Then (X_A,σ) is isomorphic to (X_B,σ) if and only if for some z, A and $B + z$ are g-equivalent.

A set A in $\mathcal{C}(G)$ will be called a Hedlund set for the generator g if for any z in G the set $\{n \in Z: z+ng \in \partial A\}$ is finite. The collection of Hedlund sets in G for g will be denoted by $\mathcal{H}(G,g)$. There are many such sets and their basic dynamical property is given by:

<u>Theorem 2</u>. The set A in $\mathcal{C}(G)$ is in $\mathcal{H}(G,g)$ if and only if whenever x and y in (X_A,σ) are proximal they are doubly asymptotic.

There are two interesting questions about these sets. First, is a Euclidean circle on the surface of a torus a Hedlund set? Do Hedlund sequences satisfy the equation

$$\lim_{n\to\infty} \frac{P(n+1,x)}{P(n,x)} = 1$$

where $P(n,x)$ is the number of n-blocks in the sequence x? (A Hedlund sequence is a characteristic sequence coming from a Hedlund set.) Note this is a generalization of the original Sturmian growth condition $P(n,x) \leq n+1$ [2].

Let x and y be characteristic sequences coming from $A \in \mathcal{C}(G)$ and $B \in \mathcal{C}(H)$ using g and h. Suppose (g,h) is a generator of $G \times H$. Then z given by $z(n) = x(n)y(n)$ is a characteristic sequence coming from $A \times B$ using (g,h). This construction seems to be basically different from $X_A \times X_B$.

<u>Theorem 3</u>. Let $A \in \mathcal{C}(K^n)$ and $B \in \mathcal{C}(K^m)$ where K denotes the circle group, and let g and h be generators of K^n and K^m such that (g,h) is a generator of K^{n+m}. If there exist $x \in \partial A$ and $y \in \partial B$ such that $x+ng \notin \partial A$ and $y+nh \notin \partial B$ for all $n \neq 0$, then $(X_{A\times B},\sigma)$ is never isomorphic to $(Y_1 \times Y_2, \phi_1 \times \phi_2)$ where (Y_i,ϕ_i) is nontrivial.

This is also a natural setting in which to extend the recent results of Furstenberg, Keynes, and Shapiro on prime flows. The following theorem generalizes their Proposition 2.1 [1] to monothetic topological groups:

<u>Theorem 4</u>. Let A be a Hedlund set and let x_1, x_2 be points in X_A. Then

$$\left\{\sum_{k=0}^{n} x_1(k) - x_2(k) : n \geq 0\right\}$$

is unbounded provided the following conditions are satisfied:

a) $p(x_1) - p(x_2) = \delta \neq 0$.

b) Let $f = \chi_A + \chi_{A+\delta}$ and let $B_i = \mathrm{Int}(f^{-1}(i))$ for $i = -1,0,1$. For at least one i, $\mathrm{Cl}(B_i) + \alpha = \mathrm{Cl}(B_i)$ implies $\alpha = 0$.

c) There exists $\gamma \in G$ such that f is not continuous at γ and

$$\{\gamma + ng : n \in Z\} \cap (\partial A \cup \partial A + \delta) = \{\gamma\}.$$

Using this theorem one can build prime flows starting with the p-adics instead of the circle. If we try to use an n-dimensional torus, $n > 1$, we can construct flows which are almost prime in the sense that a closed invariant equivalence relation must consist entirely of doubly asymptotic points.

BIBLIOGRAPHY

1. Furstenberg, Keynes, and Shapiro, Prime flows in Topological dynamics, preprint.

2. Morse and Hedlund, Symbolic dynamics II. Sturmian trajectories, Amer. J. of Math., 62(1940), 1-42.

University of Maryland

RELATIVE EQUICONTINUITY AND ITS VARIATIONS

D. McMAHON AND T. S. WU

CASE WESTERN RESERVE UNIVERSITY

The main aims of this paper concern relative equicontinuity (i.e. almost periodic extensions) and some variations. Motivated by recent results on locally almost periodic minimal sets and proximally equicontinuous minimal sets, we shall generalize these concepts to the relative case. One of the most interesting properties of a locally almost periodic minimal set is that each proximal cell shrinks to a single point under suitable motions. We ask the question: let φ be a proximal homomorphism from (X,T) onto (Y,T), where (X,T), (Y,T) are minimal sets; what is the characterization of φ so that each fiber shrinks to a singleton under suitable motions. This will be done in Section 1, where definitions of relative almost equicontinuity and relative weak almost equicontinuity are given. Most results on local almost periodicity and proximal equicontinuity are generalized.

Inspired by the work of Veech on point-distal flows and the work of Ellis and Keynes on the equicontinuous structure relation, we shall study the almost periodic extensions of minimal sets. More precisely, let φ be a homomorphism form (X,T) onto (Y,T), we ask when there exists a minimal set (Z,T) such that (Z,T) is an almost periodic extension of (Y,T) and the following diagram commutes

$$\begin{array}{ccc} (X,T) & \searrow & \\ \downarrow & & (Z,T) \\ (Y,T) & \swarrow & \end{array}$$

. In Section 2, we give some sufficient conditions on φ so that (Z,T) will be non-trivial.

While studying the above problem, we found that the openness of

φ played a critical role. In Section 3, we show that given $\varphi: (X,T) \longrightarrow (Y,T)$, there exists a homomorphism $\varphi': (X',T) \rightarrow (Y',T)$ such that φ is open, (X',T) and (Y',T) are proximal extensions of (X,T) and (Y,T) respectively, and the following diagram commutes

$$\begin{array}{ccc} X & \longleftarrow & X' \\ \varphi\downarrow & & \downarrow \varphi' \\ Y & \longleftarrow & Y' \end{array} .$$

This note is a summary and only lists the major results, the details will appear elsewhere.

1. Almost equicontinuous and weak almost equicontinuous extensions. In this section, $\varphi: (X,T) \longrightarrow (Y,T)$ is an onto homomorphism and $R = R(\varphi) = R_\varphi = \{(x,x') : \varphi(x) = \varphi(x'),\ x, x' \in X\}$. Let $\mathcal{U}\mathcal{U}_X$ be the uniformity on X and N_x the neighborhood filter at x. $P(\varphi) = P_X \cap R$ is the relativized proximal relation, P_X the proximal relation on X. $Q(\varphi) = \bigcap \{cls(\alpha T \cap R) : \alpha \in \mathcal{U}_X\}$ is the relativized regionally proximal relation [3].

(1.1) Definition. φ is almost periodic if $Q(\varphi) = \Delta$.

(1.2) Definition. φ is weakly almost equicontinuous if given x and given an index γ on X there exist $V \in N_x$ and a compact subset K of T such that $(y_1,y_2) \in R$, $y_1,y_2 \in V$ implies there exists $A \subseteq T$, $AK = T$ and $(y_1,y_2)A \subseteq \gamma$.

(1.3) Proposition. $P(\varphi) = Q(\varphi)$ iff φ is weakly almost equicontinuous.

(1.4) Definition. φ is almost equicontinuous at x if given $\gamma \in \mathcal{U}_X$, there exist $V \in N_x$ and a compact set K such that for $x' \in V$ there exists $A \subseteq T$ with $AK = T$ and $(V \cap R(x'), x')A \subset \gamma$.

(1.5) Proposition. Let (X,T) be minimal and φ in almost equicontinuous. Then given $x \in X$ and U open, there exists $t \in T$ with $P(\varphi)(x)\,t \subseteq U$.

(1.6) Proposition. Let (X,T) be minimal and be an almost equicontinuous, proximal extension of an almost periodic extension, i.e.

$$\begin{array}{ccc} (X,T) & \xrightarrow{\ \psi\ } & \\ \varphi\downarrow & & (Z,T) \\ (Y,T) & \xleftarrow{\ \pi\ } & \end{array}$$

where ψ is almost equicontinuous and proximal and π is almost periodic. Then φ is weakly almost equicontinuous, $P(\varphi) = P(\psi) = Q(\psi) = Q(\varphi)$, and given $x \in X$ and $U \in N_x$, there exists $t \in T$ with $P(\varphi)(x)\, t \subseteq U$.

(1.7) Proposition. Let (X,T) be minimal and weakly almost equicontinuous. Suppose that given U open there exists $x \in X$ with $P(\varphi)(x) \subseteq U$. Consider

$$\begin{array}{ccc} (X,T) & \xrightarrow{\ \psi\ } & \\ \varphi\downarrow & & (X/P(\varphi),T) \\ (Y,T) & \xleftarrow{\ \pi\ } & \end{array}$$

Then ψ is almost equicontinuous and proximal and π is almost periodic.

2. Relative regionally proximal relations. Let φ be a homomorphism from a minimal set (X,T) onto (Y,T). Let $S(\varphi)$ be the equicontinuous structure relation relative to φ. Then $S(\varphi)$ is the least closed equivalence relation on X such that $(X/S(\varphi),T)$ is an almost periodic extension on (Y,T). In case $Q(\varphi)$ is an equivalence relation, it is known that $Q(\varphi) = S(\varphi)$. In this section, we characterize $Q(\varphi)$ and $S(\varphi)$ under certain conditions. Now let y_0 be a fixed point of Y and x_0 a fixed point of X with $\varphi(x_0) = y_0$, let I be a minimal ideal of X, let u be an idempotent in I such that $y_0 u = y_0$, and let J be the set of all idempotents in I. Then $\{p \in I : y_0 = y_0 p\} = FJ'$, where F is a subgroup of $G = Iu$, and $J' \subseteq J$.

(2.1) Proposition. R contains a dense subset of almost periodic points if and only if $(x_0, x_0 F)T$ is dense in R.

(2.2) Proposition. If R contains a dense subset of almost periodic points, then for any $(x_0, x) \in Q(\varphi) \cap I_0 u$ there exist nets t_n, s_n in T and f_n in F such that $(u, f_n)t_n \longrightarrow (u,u)$, $(u, f_n)s_n \longrightarrow (u, p')$, and $x_0 p' = x$.

We may consider the left action of the discrete group F acting on the space I_0. Let $\hat{I}$ be a minimal set under this left action and let $\hat{X} = \varphi(\hat{I})$. Note $\hat{I} = F\hat{J}$, $\hat{J} \subseteq J$. We may assume that $u \in \hat{J}$.

(2.3) Theorem. If R contains a dense subset of almost periodic points, then $S(\varphi) = P(\varphi)Q(\varphi)$ and $S(\varphi) \cap \hat{X} \times \hat{X} = Q(\varphi) \cap \hat{X} \times \hat{X}$ and for $x \in \hat{X}$ has the form $S(\varphi)(x) \cap \hat{X} = \{x' : \text{there exist nets } r_n$ in $\hat{X}$ and t_n in T such that $(x, r_n) \longrightarrow (x, x')$, $(x, r_n)t_n \longrightarrow (x,x)\}$.

(2.4) Proposition. If X contains a relative distal point and φ is an open map, then R contains a dense subset of almost periodic points. If X is also metrizable, then $S(\varphi) \neq R$. *

(2.5) Theorem. Assume that there exists a point x in X whose relative proximal cell is finite, i.e. $P(\varphi)(x)$ is finite, and assume that φ is open, then R contains a dense subset of almost periodic points. If X is also metrizable, then $S(\varphi) \neq R$.

(2.6) Corollary. If (X,T) has a point with finite proximal cell and if X is metric, then (X,T) has a distal point.

---

* The results on the metrizable case actually were obtained by R. Ellis. At this conference, we learned that R. Ellis has obtained some results quite similar to ours. After hearing what his results were, we found that with a slight modification of our methods we may reach the same conclusions. Previously, we had to assume that $\varphi^{-1}(y_0)$ contained a dense subset of relative distal points as well as openness.

(2.7) Theorem. If there exists a dense set of relatively distal points in X_0 and if φ is open, then $Q(\varphi) \cap X_0 \times X_0 = S(\varphi) \cap X_0 \times X_0$ and has the form $S(\varphi)(x) \cap X_0 = \{x' :$ there exist nets r_n in X_0 and t_n in T such that $(x,r_n) \longrightarrow (x,x'),\ (x,r_n)t_n \longrightarrow (x,x)\}$ for x in X_0; also $S(\varphi) = P(\varphi)Q(\varphi)$.

3. Openness of homomorphisms. In view of the results in Section 2, we see that φ being open plays a critical role, in our judgement. Here we give a theorem on this aspect.

(3.1) Theorem. Let (X,T) be minimal and φ be a homomorphism of (X,T) onto (Y,T). Then there exist minimal sets (X',T) and (Y',T) and a homomorphism $\varphi' : (X',T) \longrightarrow (Y',T)$ such that φ' is open, (X',T) and (Y',T) are proximal extensions of (X,T) and (Y,T) respectively, and the following diagram commutes

$$\begin{array}{ccc} X & \longleftarrow & X' \\ \varphi \downarrow & & \downarrow \varphi' \\ Y & \longleftarrow & Y' \end{array}.$$

APPROXIMATION BY MEASURE-PRESERVING HOMEOMORPHISMS

John C. Oxtoby

Bryn Mawr College

1. Statement of Theorem. Let X be a metrizable topological space with a connected open subset M each point of which has a neighborhood homeomorphic to Euclidean r-space E^r, where $r \geqq 2$. Let μ be a non-atomic, normalized Borel measure in X such that $\mu(M) = 1$ and $\mu(G) > 0$ for every non-empty open set $G \subset M$. Alternatively, μ may be taken to be the completion of such a measure.

Theorem 1. If T is an invertible, measurable, and μ-preserving transformation of X (that is, T: X → X is 1-1, onto, and $\mu(T^{-1}E) = \mu(E) = \mu(TE)$ whenever E is μ-measurable), then for each $\varepsilon > 0$ there exists a μ-preserving homeomorphism S of X onto itself such that S is equal to T except on a set of measure less than ε, and S is equal to the identity outside some compact r-cell contained in M.

2. Preliminary remarks. Let T' be equal to T on the set $\bigcap_{n=-\infty}^{\infty} T^n M$ and equal to the identity elsewhere in X. Then T' is equal to T almost everywhere and T'M = M. The subspace M is topologically complete and separable (in fact, locally compact). Hence, in case μ is not a Borel measure but the completion of one, a well-known theorem of von Neumann [6] implies that T' is equal almost everywhere to a 1-1 Borel measurable transformation T". If the theorem holds for T" and the Borel restriction of μ, then it holds for T. Consequently we need consider only Borel measures and transformations, and we may assume that TM = M. It is obvious that the theorem holds for X if it holds for the subspace M.

Secondly, let X' denote the one-point compactification of M, let μ' be the (unique) nonatomic Borel measure in X' that coincides with μ on M, and let T' denote the (unique) transformation of X' that is equal to T on M. Then X', M, μ', T' satisfy the hypotheses of the theorem, and X' is a compact metrizable space. If the conclusion of the theorem holds for this space, then it holds also for X. Consequently it is sufficient to prove the theorem when X is a compact metric space.

Theorem 1 may be interpreted to mean that in the group of all invertible, measurable, and μ-preserving transformations of X with the uniform topology defined by Halmos [3], the μ-preserving homeomorphisms (or rather, the equivalence classes that include at least one homeomorphism) constitute a dense subset.

3. Lemmas concerning the space X. A subset σ of a topological space is called an r-cell if σ is a topological image of the closed unit ball B in E^r. Its boundary is denoted by $\partial\sigma$ and its interior by int σ. The only r-cells we shall consider will be subsets of M or of E^r. In either case it follows from invariance of region that if f is a homeomorphism of B onto σ, then int σ = f(int B) and $\partial\sigma = f(\partial B)$. Let p = f(0), where 0 is the center of B. For any t > 1 let φ^t denote the radial contraction of B that maps each point P of B at distance ρ from 0 onto the point of 0P at distance ρ^t from 0. Let g^t be equal to $f\varphi^t f^{-1}$ on σ and equal to the identity on the complement of σ in X (or E^r). Then g^t is a homeomorphism of X (or E^r) onto itself. We shall call such a mapping g^t a contraction of σ toward p. If K is a compact subset of int σ and U is a neighborhood of p, then $g^t(K) \subset U$ for all sufficiently large t. When h is a homeomorphism of X (or E^r) onto itself, the set supp h = $\{x : h(x) \neq x\}$ is called the support of h. Thus supp g^t = int σ - $\{p\}$.

Assume now that X is compact. Let H[X] denote the group of homeomorphisms of the metric space (X,d) onto itself, and let

$$H_M[X] = \{h \in H[X] : \text{supp } h \subset M\}.$$

Thus $H_M[X]$ is the set of homeomorphisms of X that are equal to the identity on X - M. Both H[X] and $H_M[X]$ are topologically complete metric groups with respect to the metric of uniform convergence:

$$\rho(h,g) = \max_{x \in X} d(h(x),g(x)).$$

The following lemma (or rather its corollary) is a generalization of Theorem 1 of [8].

<u>Lemma 1.</u> <u>If</u> X <u>is compact and</u> F <u>is a nowhere dense closed subset of</u> X, <u>then the set</u> $\{h \in H_M[X] : \mu(hF) = 0\}$ <u>is a dense</u> G_δ <u>subset of</u> $H_M[X]$.

<u>Proof.</u> For any $\varepsilon>0$ let $\mathcal{E} = \{h \in H_M[X] : \mu(hF) < \varepsilon\}$. It is evidently sufficient to show that $\mathcal{E}$ is open and dense in $H_M[X]$. If $h \in \mathcal{E}$, then hF is contained in an open set G with $\mu(G) < \varepsilon$, and $gF \subset G$ for all g sufficiently near h. Hence $\mathcal{E}$ is open. Since hF is also nowhere dense and closed in X, to show that $\mathcal{E}$ is dense in $H_M[X]$ it is sufficient to show that it always has members arbitrarily near the identity.

Since μ is regular there exists a compact set $K \subset F \cap M$ and an open set $G \supset F$ such that $\mu(G - K) < \varepsilon/3$. Let $\delta > 0$ be given. Each point of K belongs to the interior of an r-cell τ contained in $G \cap M$ with $\mu(\partial\tau) = 0$. Hence there exists a finite collection of such r-cells $\tau_1, \cdots, \tau_k$ whose interiors cover K. The sets

$$K \cap (\tau_i - \bigcup_{j<i} \tau_j) \ (i=1,\cdots,k)$$

partition K, and so there exist disjoint compact sets $K_1, \cdots, K_k$ (possibly empty) such that

$$K_i \subset K \cap (\tau_i - \bigcup_{j<i} \tau_j) \text{ and } \sum_{i=1}^{k} \mu(K_i) > \mu(K) - \varepsilon/3.$$

Let η be a positive number less that δ such that the η-neighborhoods of K_i and K_j are disjoint whenever $i \neq j$. Divide each τ_i into non-overlapping r-cells τ_{ij} $(j = 1, \cdots, \ell)$ of diameter less than η such that $\mu(\partial\tau_{ij}) = 0$. For each i and j let K_{ij} be a compact subset of $K_i \cap \operatorname{int} \tau_{ij}$ such that $\mu(K_{ij}) > \mu(K_i \cap \operatorname{int} \tau_{ij}) - \varepsilon/3k\ell$, and let σ_{ij} be an r-cell such that $K_{ij} \subset \operatorname{int} \sigma_{ij} \subset \sigma_{ij} \subset \operatorname{int} \tau_{ij}$. Put $K_0 = \bigcup_{i=1}^{k} \bigcup_{j=1}^{\ell} K_{ij}$. Then $\mu(K_0) = \sum_{i=1}^{k} \sum_{j=1}^{\ell} \mu(K_{ij}) > \sum_{i=1}^{k} [\mu(K_i) - \varepsilon/3k] > \mu(K) - 2\varepsilon/3$. Hence $\mu(G - K_0) < \varepsilon$, and

$\{\sigma_{ij} : 1 \leq i \leq k, 1 \leq j \leq \ell\}$ is a finite collection of disjoint r-cells contained in $G \cap M$. For each i and j we can choose a point $p_{ij} \in \text{int}\, \sigma_{ij} - F$, since F is nowhere dense. Let g_{ij} be a contraction of σ_{ij} toward p_{ij} such that $g_{ij} K_{ij} \subset \text{int}\, \sigma_{ij} - F$, and let g denote the product of all the homeomorphisms g_{ij}. Then $\text{supp}\, g \subset M$ and $g(K_o) \subset G - F$. Consequently $g^{-1}(F) \subset G - K_o$. Hence $\mu(g^{-1}F) < \varepsilon$, and therefore g^{-1} belongs to $\mathcal{E}$. Since g^{-1} only moves points within the cells σ_{ij} its distance from the identity is less than δ.

<u>Corollary. When X is compact and A is a subset of X such that $A \cap M$ is of first category, there exists an $h \in H_M[X]$ such that hA is contained in a set of μ-measure zero. Such homeomorphisms constitute a residual set in $H_M[X]$.</u>

The following lemma was suggested by a remark of Katok and Stepin [4].

<u>Lemma 2.</u> M <u>always contains an open set P homeomorphic to E^r such that</u> $\mu(P) = 1$.

<u>Proof.</u> M is an open r-dimensional manifold. By a theorem of Doyle and Hocking [2], M contains an open set Q homeomorphic to E^r such that M - Q is closed and nowhere dense relative to M. Imbed M in a compact space X. Then $F = \text{cl}\,(M - Q)$ is a nowhere dense closed subset of X. By Lemma 1 there exists a homeomorphism h with $\text{supp}\, h \subset M$ such that $\mu(hF) = 0$. Therefore $P = hQ$ is a subset of M homeomorphic to E^r, and $\mu(P) = 1$.

Thus M can always be taken homeomorphic to E^r. The assumption that X is compact has now served its purpose. (It was needed to facilitate consideration of H[X].) Dropping this restriction we see that it is sufficient to prove Theorem 1 under the assumption that X itself is homeomorphic to E^r.

4. Topologically equivalent measures in the r-dimensional cube.

Let R_o denote the interior of an r-dimensional cube R in E^r, $r \geqq 2$. Let $\mathcal{M}[R_o]$ denote the class of nonatomic, finite Borel measures in R_o that are positive on every non-empty open subset of R_o. If $\mu \in \mathcal{M}$, h is a homeomorphism of R_o onto itself, and ν is defined by $\nu(E) = \mu(hE)$ for every Borel set E, then $\nu \in \mathcal{M}$ and $\nu(R_o) = \mu(R_o)$. Conversely, it is known that if μ and ν belong to $\mathcal{M}$ and $\nu(R_o) = \mu(R_o)$, then there exists a homeomorphism h of E^r with supp $h \subset R_o$ such that $\nu(E) = \mu(hE)$ for every Borel set E ([9] Theorem 2; see also [4]). It is to be emphasized that the transformation of μ into ν can always be effected by a homeomorphism that reduces to the identity on ∂R. The following corollary of this theorem is also known.

Lemma 3 ([9] Corollary 4). *If μ and ν belong to $\mathcal{M}[R_o]$, $\mu(R_o) = \nu(R_o)$, and L is a straight line segment contained in R_o such that $\mu(L) = \nu(L) = 0$, then there exists a homeomorphism h of E^r with supp $h \subset R_o - L$ such that $\nu(E) = \mu(hE)$ for every Borel set $E \subset R_o$.*

The proof (loc. cit.) consists in defining a continuous map f of R onto itself, by identifying certain boundary points, such that $f(\partial R) \supset L$ and such that the restriction of f to R_o is a homeomorphism. It will not be repeated here. For present purposes we need a further corollary.

Lemma 4. *If μ and ν belong to $\mathcal{M}[R_o]$ and $\mu(R_o) = \nu(R_o)$, if L is a straight line segment contained in R_o, and if C is a Cantor set such that $C \subset L$ and $\mu(C) = \nu(C) = 0$, then there exists a homeomorphism h of E^r with supp $h \subset R_o - C$ such that $\nu(E) = \mu(hE)$ for every Borel set $E \subset R_o$.*

Proof. Take L to be the shortest line segment containing C. Let J_i ($i = 1, 2, \cdots$) be the closures of the components of L - C. For each i let σ_i be a rectangular r-cell contained in R_o having J_i for its longest diameter. Then the cells σ_i are pairwise disjoint. Choose $p_i \in \text{int}\, \sigma_i - J_i$ and let g_i^t be a contraction of σ_i toward p_i that leaves invariant each ray from p_i. Let g^t denote the product of all the g_i^t $(i = 1, 2, \cdots)$. Then g^t is a homeomorphism with $\text{supp}\, g^t \subset R_o - C$ and the sets $g^t(L - C)$ for different values of t are mutually disjoint. Hence we can fix t so that $\mu(g^t(L - C)) = \nu(g^t(L - C)) = 0$. Define $\mu'(E) = \mu(g^tE)$ and $\nu'(E) = \nu(g^tE)$ for each Borel set $E \subset R_o$. Then μ' and ν' belong to $\mathcal{M}[R_o]$, $\mu'(R_o) = \nu'(R_o)$, and $\mu'(L) = \nu'(L) = 0$. By Lemma 3 there exists a homeomorphism g with $\text{supp}\, g \subset R_o - L$ such that $\nu'(E) = \mu'(gE)$ for every Borel set $E \subset R_o$. Then $h = g^t g (g^t)^{-1}$ is a homeomorphism with $\text{supp}\, h \subset R_o - C$ such that $\nu(E) = \mu(hE)$ for every Borel set $E \subset R_o$.

Remark. Lemma 4 still holds when the hypothesis "L is a straight line segment" is replaced by "L is the union of a finite number of disjoint straight line segments."

Proof. It is clear that any such set L can be mapped onto a subset of a single line segment by a homeomorphism g with $\text{supp}\, g \subset R_o$. Let $C' = gC$, $\mu'(E) = \mu(g^{-1}E)$, and $\nu'(E) = \nu(g^{-1}E)$. Then μ', ν', and C' satisfy the hypotheses of Lemma 4, and so $\nu'(E) = \mu'(h'E)$ for some h' with $\text{supp}\, h' \subset R_o - C'$. It follows that the homeomorphism $h = g^{-1} h' g$ satisfies the conclusion of Lemma 4.

5. Extension of homeomorphisms between Cantor sets. Let C and C' be Cantor sets in E^r. When $r = 2$ it was shown by Antoine [1] that any homeomorphism of C onto C' can be extended to a homeomorphism of E^r, but that when $r \geqq 3$ such an extension may be impossible. However, it was shown by Keldyš [5] that if C and C' are "cellularly separated," or equivalently, if each can be mapped into a subspace E^1 of E^r by a homeomorphism of E^r, then any homeomorphism of C onto C' can be extended to a homeomorphism of E^r that is isotopic to the identity. (I am

indebted to K. Krigelman for calling my attention to this theorem.) For our purposes it will be sufficient to consider Cantor sets C and C' that can both be covered by a finite number of disjoint straight line segments, but we need a stronger conclusion, namely that the extension can be taken equal to the identity outside a given domain that contains these line segments. The methods used by Keldyš actually establish this conclusion. Nevertheless we prefer to follow a different line of reasoning.

Lemma 5. Let C and C' be Cantor sets contained in the union K of a finite number of disjoint straight line segments in E^r, $r \geq 1$, and let U be a connected open set that contains K. There exists a homeomorphism h of E^r with $\operatorname{supp} h \subset U$ such that $hC = C'$.

Proof. By induction on the number of line segments it is easy to see that K can be mapped onto a subset of one of its components. Hence we may assume that C and C' are both contained in a single straight line segment L, and that the endpoints of L do not belong to either C or C'. By a suitable choice of axes we can assume that L is the interval $[-1, 1]$ of the x_1-axis. Let the ε-neighborhood of L be contained in U. The intervals I_i and I_i' composing L - C and L - C', respectively, can be numbered in such a way that I_i lies to the left of I_j if and only if I_i' lies to the left of I_j'. Let f map each I_i linearly onto I_i'. Extend f over C by continuity and let $f(x) = x$ for $|x| \geq 1$. Then f is a homeomorphism of E^1, with $\operatorname{supp} f \subset (-1, 1)$, that maps C onto C'. If $r > 1$, extend f to E^2 by projecting f from the points $(0, \pm\varepsilon)$, that is, let h_2 be the homeomorphism of the xy-plane defined by

$$h_2(x,y) = (g(x,y), y), \text{ where } \quad g(x,y) = \begin{cases} \left(1 - \frac{|y|}{\varepsilon}\right) f\left(\frac{\varepsilon x}{\varepsilon - |y|}\right) & \text{when } |y| < \varepsilon \\ x & \text{when } |y| \geq \varepsilon. \end{cases}$$

For $2 < i \leq r$ define h_i by projecting h_{i-1} from the points with coordinates $\pm\varepsilon$ on the x_i-axis. Then $h = h_r$ will be a homeomorphism with $\operatorname{supp} h \subset U$ that maps C onto C'.

<u>Lemma 6.</u> <u>Let</u> C <u>and</u> C' <u>be Cantor sets contained in the union</u> K <u>of a finite number of disjoint straight line segments in</u> E^r, $r \geq 2$, <u>let</u> U <u>be a connected open set that contains</u> K, <u>and let</u> φ <u>be a homeomorphism of</u> C <u>onto</u> C'. <u>There exists a homeomorphism</u> h <u>of</u> E^r <u>with</u> $\operatorname{supp} h \subset U$ <u>such that</u> h <u>is an extension of</u> φ.

<u>Proof.</u> Let C" be the ordinary Cantor ternary set constructed on a straight line segment L contained in U - K. By Lemma 5 there exist homeomorphisms g_1 and g_2, supported by U, such that $g_1C = C'' = g_2C'$. Let $\psi = g_2\varphi g_1^{-1}$. Then ψ is a homeomorphism of C" onto itself. If g is an extension of ψ, then $h = g_2^{-1} g\, g_1$ is an extension of φ. Hence it is sufficient to prove the lemma for the case in which C and C' coincide with the ternary set on L, which we may take to be the interval $I = [-1, 1]$ of the x_1-axis. Choose $0 < \varepsilon < 1$ so that the rectangular closed ε-neighborhood σ of I is contained in U. Instead of extending φ we shall seek an extension h of φ^{-1}.

To begin with let us confine attention to the x_1x_2-plane E^2. Let I(0) and I(1) denote the left and right thirds of I. For $i_k = 0, 1$ and $n \geq 1$ let $I(i_1, \cdots, i_n, 0)$ and $I(i_1, \cdots, i_n, 1)$ denote the left and right thirds of $I(i_1, \cdots, i_n)$, respectively. Let $\sigma(i_1, \cdots, i_n)$ be the rectangular closed $(\varepsilon/3^n)$-neighborhood of $I(i_1, \cdots, i_n)$ in E^2. Then it is clear that the 2-cells $\sigma(i_1, \cdots, i_n)$ are disjoint for each n, that $\sigma(i_1, \cdots, i_{n+1}) \subset \operatorname{int} \sigma(i_1, \cdots, i_n)$, that $\operatorname{diam} \sigma(i_1, \cdots, i_n) = 2\sqrt{(1+\varepsilon)^2 + \varepsilon^2}/3^n$, and that $C = \bigcap_{n=1}^{\infty} \bigcup_{i_1, \cdots, i_n} \sigma(i_1, \cdots, i_n)$. Define $C(i_1, \cdots, i_n) = C \cap \sigma(i_1, \cdots, i_n)$.

The compact sets $\varphi C(0)$ and $\varphi C(1)$ partition C. Hence there is a least positive integer j such that none of the intervals $I(i_1, \cdots, i_j)$ meets both $\varphi C(0)$ and $\varphi C(1)$. Therefore $\varphi C(0)$ is the union of certain of the sets $C(i_1, \cdots, i_j)$, say $c_1, \cdots, c_k$, and $\varphi C(1)$ is the union of the others, say $c_1', \cdots, c_\ell'$ where $k > 0$, $\ell > 0$, and $k + \ell = 2^j$.

Let g be a homeomorphism of E^2 with $\text{supp } g \subset \text{int } \sigma$ that translates $c_1 \cup c_2 \cup \cdots \cup c_k$ upward by $2\varepsilon/3$, and translates $c_1' \cup \cdots \cup c_\ell'$ downward by $2\varepsilon/3$. The polygon with vertices $(-1-\varepsilon, -\varepsilon/3)$, $(0, -\varepsilon/3)$, $(0, \varepsilon/3)$, $(1+\varepsilon, \varepsilon/3)$ divides σ into polygonal 2-cells σ' and σ'' such that $C(0) \cup g\,\varphi C(0) \subset \text{int } \sigma'$, $C(1) \cup g\,\varphi C(1) \subset \text{int } \sigma''$, and $\text{int } \sigma' \cap \text{int } \sigma'' = \emptyset$. By Lemma 5 there exists a homeomorphism g' with support contained in $\text{int } \sigma'$ such that $g'\,g\,\varphi C(0) = C(0)$, and a homeomorphism g'' with $\text{supp } g'' \subset \text{int } \sigma''$ such that $g''\,g\,\varphi C(1) = C(1)$. Let $h_1 = g''\,g'\,g$. Then $\text{supp } h_1 \subset \text{int } \sigma$ and $h_1 \varphi C(i_1) = C(i_1)$ for $i_1 = 0, 1$.

Replacing σ by $\sigma(i_1)$, C by $C(i_1)$, and φ by $h_1\varphi$, exactly the same reasoning leads to a homeomorphism $h(i_1)$ with $\text{supp } h(i_1) \subset \text{int } \sigma(i_1)$ such that $h(i_1)\,h_1 \varphi C(i_1, i_2) = C(i_1, i_2)$ for $i_2 = 0, 1$. Put $h_2 = h(0)h(1)h_1$. Then $h_2\,\varphi\,C(i_1, i_2) = C(i_1, i_2)$ for all i_1, i_2. By induction we obtain a sequence of homeomorphisms h_n such that

(1) $$h_n \varphi C(i_1, \cdots, i_n) = C(i_1, \cdots, i_n)$$

(2) $$h_{n+1}\, h_n^{-1} \sigma(i_1, \cdots, i_n) = \sigma(i_1, \cdots, i_n)$$

(3) $$\text{supp } h_{n+1}\, h_n^{-1} \subset \bigcup_{i_1, \cdots, i_n} \text{int } \sigma(i_1, \cdots, i_n)$$

for all $i_1, \cdots, i_n$. It follows from (1), (2), and (3), that the sequence h_n converges uniformly to a continuous mapping h, that $hp = h_n p$ for all sufficiently large n when $p \in E^2 - C$, and that $h\varphi p = p$ when $p \in C$. Consequently h is 1-1 and therefore a homeomorphism of E^2. Since $\text{supp } h \subset \text{int } \sigma$ and $h = \varphi^{-1}$ on C, this completes the proof in case $r = 2$. When $r > 2$ we need only project h successively from the points with coordinates $\pm\varepsilon$ on each of the remaining axes to obtain a homeomorphism of E^r that extends φ^{-1} and whose support is contained in U.

<u>Lemma 7. Let C and C' be Cantor sets contained in the union K of a finite number of disjoint straight line segments in the interior R_1 of</u>

<u>an r-cube, $r \geqq 2$. Let $\mu \in \mathcal{M}[R_1]$ and let φ be a μ-preserving homeomorphism of C onto C'. Then φ can be extended to a homeomorphism h of E^r with supp $h \subset R_1$ such that the restriction of h to R_1 is μ-preserving.</u>

<u>Proof.</u> By Lemma 6, φ can be extended to a homeomorphism g with supp $g \subset R_1$. Define $\nu(E) = \mu(E - C')$ and $\nu'(E) = \mu(g^{-1}E - C)$ for Borel sets $E \subset R_1$. Then ν and ν' belong to $\mathcal{M}[R_1]$. Moreover, $\nu(R_1) = \nu'(R_1)$ and $\nu(C') = \nu'(C') = 0$. By the Remark following Lemma 4 there exists a homeomorphism g' with supp $g' \subset R_1 - C'$ such that $\nu'(E) = \nu(g'E)$. Take $h = g'g$. Then h is a homeomorphism of E^r with supp $h \subset R_1$ that extends φ. For any Borel set $E \subset R_1 - C$ we have $\mu(hE) = \mu(g'gE - C') = \nu(g'gE) = \nu'(gE) = \mu(E - C) = \mu(E)$. For any Borel set $E \subset C$ we have $\mu(hE) = \mu(\varphi E) = \mu(E)$. Hence h is μ-preserving on R_1.

<u>6. Proof of Theorem 1.</u> Let R_o be the interior of the closed unit cube R in E^r. If the theorem is true for one (normalized) measure μ in $\mathcal{M}[R_o]$, then it is true for every normalized measure ν in $\mathcal{M}[R_o]$, as may be seen by considering the conjugate of T by a homeomorphism that transforms μ into ν. For present purposes it is convenient to take a special measure μ. Let $\{L_i\}$ be a sequence of disjoint straight line segments $L_i \subset R_o$ whose union is dense in R_o. Let λ_i denote linear Lebesgue measure on L_i, and define $\mu(E) = \sum_{i=1}^{\infty} \lambda_i(E \cap L_i)/2^i \lambda_i(L_i)$ for Borel sets $E \subset R_o$. Then μ belongs to $\mathcal{M}[R_o]$ and $\mu(R_o) = 1$.

Let T be a 1-1 μ-preserving transformation of R_o. By Lusin's theorem (applied to each coordinate of Tx), for any $\varepsilon > 0$ there exists a set $A \subset R_o$ with $\mu(A) > 1 - \varepsilon$ such that the restriction of T to A is continuous. Put $B = A \cap \bigcup_{i=1}^{k} L_i \cap \bigcup_{i=1}^{k} T^{-1}L_i$, where k is chosen large enough so that $\mu(B) > 1 - \varepsilon$. Then B contains a Cantor set C with $\mu(C) > 1 - \varepsilon$ (see, for example, [7]). The restriction φ of T to C is 1-1 and continuous. Hence φ is a μ-preserving homeomorphism of C onto C' = TC. Both C and C' are contained in $K = L_1 \cup \cdots \cup L_k$. Let R_1 be the interior of a cube concentric with R such that

$$K \subset R_1 \subset \mathrm{cl}\, R_1 \subset R_0,$$

and let μ_1 denote the restriction of μ to subsets of R_1. Then μ_1 belongs to $\mathfrak{M}[R_1]$ and φ is a μ_1-preserving homeomorphism of C onto C'. By Lemma 7, φ can be extended to a homeomorphism h of E^r with supp $h \subset R_1$ such that the restriction of h to R_1 is μ_1-preserving. Let S denote the restriction of h to R_0. Then S is a μ-preserving homeomorphism of R_0 onto itself, S is equal to the identity outside the r-cell cl R_1, S is equal to T on C, and $\mu(C) > 1 - \varepsilon$. It follows that the theorem is true when $X = R_0$ and μ is any member of $\mathfrak{M}[R_0]$ with $\mu(R_0) = 1$. We have already remarked (following Lemma 2) that this is sufficient to establish Theorem 1 in all generality.

<u>Added Note.</u> While this paper was being typed I received from H. E. White, Jr. a preprint of a paper by him entitled "The approximation of one-one measurable transformations by measure preserving homeomorphisms." In this paper he gives a different proof of Theorem 1 in the case of the open unit cube I^r. (The last clause of Theorem 1 is not explicitly stated but could be.) His proof is based on [9] and on a paper by C. Goffman entitled "One-one measurable transformations" in Acta Math. 89, 261-278 (1953), which treats the approximation of one-one measurable transformations by homeomorphisms (not measure-preserving) by methods that have much in common with ours. See also White's paper "The approximation of one-one measurable transformations by diffeomorphisms," Trans. Amer. Math. Soc. 141, 305-322 (1969).

REFERENCES

[1] Antoine, L., Sur l'homéomorphie de deux figures et de leurs voisinages, J. Math. Pures Appl. (8) 4, 221-325 (1921).

[2] Doyle, P. H. and Hocking, J. G., A decomposition theorem for n-dimensional manifolds, Proc. Amer. Math. Soc. 13, 469-471 (1962).

[3] Halmos, P. R., Lectures on ergodic theory, Chelsea, New York, 1956.

[4] Katok, A. B. and Stepin, A. M., Metric properties of measure preserving homeomorphisms (Russian), Uspehi Mat. Nauk 25, no. 2 (152), 193-220 (1970) (Russian Math. Surveys 25, 191-220 (1970)).

[5] Keldyš, L. V., Topological imbeddings in Euclidean space, Proc. Steklov Inst. Math. No. 81 (1966) (translation Amer. Math. Soc. 1968).

[6] Neumann, J. v., Einige Sätze über messbare Abbildungen, Ann. of Math. (2) 33, 574-586 (1932).

[7] Oxtoby, J. C., Homeomorphic measures in metric spaces, Proc. Amer. Math. Soc. 24, 419-423 (1970).

[8] Oxtoby, J. C. and Ulam, S. M., On the equivalence of any set of first category to a set of measure zero, Fund. Math. 31, 201-206 (1938).

[9] Oxtoby, J. C. and Ulam, S. M., Measure-preserving homeomorphisms and metrical transitivity, Ann. of Math. (2) 42, 874-920 (1941).

CLASS PROPERTIES OF DYNAMICAL SYSTEMS

WILLIAM PARRY

UNIVERSITY OF WARWICK

§0. Introduction.

By a dynamical system I shall mean a group S of measure preserving transformations of a Lebesgue space $(X, \mathcal{B}, m)$ or of homeomorphisms of a compact metric space X. S will usually be parameterized by R or Z. In other words, the object of study is a representation in the group of all measure preserving transformations or of all homeomorphisms, in contrast with a representation in the unitary group of a Hilbert space. In the latter case one proceeds to study representations through related simple representations e.g. irreducible representations. When sufficient information concerning "component" representations is at hand, one proceeds to rebuild the original representation from these components. The situation is sometimes similar in the case of dynamical systems, although irreducibles, perhaps, have not the same significance. Certainly when S is abelian, ergodic and has discrete spectrum, we see that S has sufficiently many representations as translations on a circle or finite cyclic group. The conclusion to Halmos-Von-Neumann's study [5] in this area is that S has a faithful representation in terms of translations of a compact abelian group. Similar resarks apply to non-abelian groups with discrete spectra in the sense of Mackey [6] (the associated unitary representation should decompose into finite dimensional representations.) The conclusion here is that S has a faithful representation in terms of translations of a quotient space of a compact group.

This point of view concerning dynamical systems requires a class of dynamical systems $\mathcal{C}$ which, if not simple enough, is at least reasonably well understood. Dynamical systems with discrete spectra are studied via the class $\mathcal{C}$ of systems which arise from translations on homogeneous spaces of compact Lie groups or of $U(n)$. If S is abelian then $\mathcal{C}$ may be reduced to systems consisting of translations on direct products of cyclic groups with torii. If S has sufficiently many such simple representations the problem is then to rebuild S from these representations in terms of projective limits.

For the sake of concreteness let us suppose that $S = Z$ (the group of integers) represented as the iterates of a single transformation T (invertible measure preserving transformation or homeomorphism.)

If T_1, T_2 are two such transformations we say that T_2 is a factor of T_1, if there is an *allowable* map F such that $FT_1 = T_2F$ written $T_1 \xrightarrow{F} T_2$. By allowable we mean F is surjective and measure preserving when T_1, T_2 are measure preserving and continuous when T_1, T_2 are homeomorphisms. Of course other restrictions on F may be imposed according to the problem under consideration. If F is invertible and F^{-1} is allowable then T_1 and T_2 are *isomorphic* via the *isomorphism* F. Sometimes we will say that F represents T_1 as T_2 and does so *faithfully* when F is an isomorphism.
All transformations will be assumed "sufficiently ergodic" i.e. ergodic, totally ergodic, minimal or topologically transitive depending on the problem.

We shall be interested in the following properties of a class $\mathcal{C}$ of *model* transformations:

A. If $$\begin{array}{ccc} & T & \\ {\scriptstyle F_1}\swarrow & & \searrow{\scriptstyle F_2} \\ T_1 & & T_2 \end{array}$$ where $T_1, T_2 \in \mathcal{C}$ and F_1, F_2 separate T, then T is isomorphic to a member of $\mathcal{C}$.

B. If T_1, $T_2 \in \mathcal{C}$ and $T_1 \xrightarrow{F} T_2$ then F is "known".

C. If $T \in \mathcal{C}$ and $T \xrightarrow{F} S$ then S is isomorphic to a member of $\mathcal{C}$.

D. If T is a projective limit of members of $\mathcal{C}$ then T is isomorphic to a member of $\mathcal{C}$.

B. is vague but means that complete knowledge of T_1, T_2 leads to a determination of the set of such F.

The significance of C is not clear although it looks like a desirable property.

In some sense D is not serious as we could form the class $\overline{\mathcal{C}}$ of inverse limits of members of $\mathcal{C}$ but then $\overline{\mathcal{C}}$ may not enjoy C or B. In fact this problem needs further study.

A. is the most important property as it acts as an organizing lemma for representing transformations as inverse limits. By separating we mean that F_1, F_2 separate points in the topological context or that the σ-algebra associated with T is generated by the F_1, F_2 pull-backs of the σ-algebras associated with T_1, T_2, in the measure theoretic context.

§1. Shifts of finite type.

T is called a <u>shift of finite type</u> or intrinsic Markov chain, if T is the restriction to $X \subset \prod_{Z} \{1,2,\dots k\}$ of the shift transformation where X is a closed shift invariant set and where X is determined by a finite set of words $W_1, \dots W_m$ i.e. $x \in X$ if and only if each of its words is $W_1, W_2, \dots$ or W_m. Let $\mathcal{C}$ be the class of topologically transitive shifts of finite type. F is allowable if F is continuous, surjective, and finite to one. In this case:

$\mathcal{C}$ satisfies A, B and C. B is the only property of any depth here and the form of allowable maps is contained in William's theory [10]. Essentially, they are codings of a very simple form. What do the projective limits of elements of $\mathcal{C}$ look like? Does $\overline{\mathcal{C}}$ enjoy the main properties of shifts of finite type?

If $\mathcal{C}$ is taken to be the class of Anosov diffeomorphisms, how should "allowable" be defined so that A, B are satisfied? In this connection if T is the Anosov diffeomorphism defined by $\begin{pmatrix} 2 & 1 \\ 1 & 1 \end{pmatrix}$ on the (additive) torus X and $(x,y) \longrightarrow (-x,-y)$ defines a Z_2 action, then the factor of T on X/Z_2 (a 2-sphere) is not an Anosov diffeomorphism.

§2. Generalized discrete spectrum and the Klein bottle.

Let T be a minimal homeomorphism of a compact metric space X and let $\mathcal{A}$ be a closed T invariant * sub-algebra of $C(X)$. Let $G(\mathcal{A}) = \{f \in C(X): |f| = 1, \ fT/f \in \mathcal{A}\}$ and let $D(\mathcal{A})$ be the closure in $C(X)$ of the linear span of $G(\mathcal{A})$. $D^n(\mathcal{A}) = D(D^{n-1}\mathcal{A})$ where $D^0(\mathcal{A}) = \mathcal{A}$. T is said to have generalized discrete spectrum (of finite type) if $D^n(\mathbb{C}) = C(X)$ for some positive integer n where $\mathbb{C}$ is the algebra of constant functions.

Let $\mathcal{C}$ be the class of minimal homeomorphisms with generalized discrete spectra. In [4] it was shown that members of $\mathcal{C}$ can exist only on compact abelian groups. It is an easy matter to show that $\mathcal{C}$ enjoys A. I have no comment to make on B, but $\mathcal{C}$ does not satisfy C.

Let $T(x,y) = (ax, \phi(x)y)$ (multiplicative notation) be a minimal homeomorphism of the torus $K \times K$, where a is not of finite order and ϕ is a suitable continuous map of K to K. Let $g(x,y) = (-x, y^{-1})$ so that g^2 is the identity, and $Z_2 = (e,g)$. $K \times K/Z_2$ is the Klein bottle and $g(ax, \phi(x)y) = (-ax, \phi(x)^{-1} y^{-1})$ and $Tg(x,y) = T(-x, y^{-1}) = (-ax, \phi(-x)y^{-1})$ so that $Tg = gT$ if $\phi(-x)\phi(x) = 1$.

The fact is such ϕ can be found yielding a minimal factor transformation T_1 on the Klein bottle so that T_1 cannot have generalized discrete spectrum, although it is clear that T has.

§3. Bernoulli transformations

Let $\mathcal{C}$ be the class of Bernoulli transformations. Ornstein has shown that $\mathcal{C}$ satisfies C and D [7]. If ergodic transformations which are not mixing are allowed then $\mathcal{C}$ does not satisfy A as can be shown using the example of a random walk on the vertices of a square which moves one step to an adjacent vertex with probability $\frac{1}{2}$. (Ken Thomas pointed this out to me.) This gives us

$$\begin{array}{ccc} & T & \\ F_1 \swarrow & & \searrow F_2 \\ T_1 & & T_2 \end{array}$$

where F_1, F_2 are separating and T_1, T_2 are the Bernoulli $(\frac{1}{2}, \frac{1}{2})$ shifts. T is not Bernoulli because it is not mixing. The serious question is whether we can assert that T is Bernoulli when T is assumed to be a K-automorphism and as far as I am aware this problem has not been solved. If the answer is affirmative one could speak of maximum Bernoulli factors of K automorphisms.

§4. Quasi-discrete spectra.

Without defining this term here (c.f. Abramov [1]) I should mention that Hahn and I [3] identified this class in terms of representations as unipotent affines on torii. The details are completed as soon as the class $\mathcal{C}$ of ergodic unipotent affines is fully analysed.

An affine transformation of one torus to another is a composition of an endomorphism and a translation. If the endomorphism is an isomorphism then the affine is invertible. If the two torii coincide and the automorphism part of the affine is unipotent (i.e. its matrix form is unipotent) then the affine is called unipotent.

The class $\mathcal{C}$ of ergodic unipotent affines satisfies A, B, C in both topological and measure-theoretic contexts. A totally ergodic (minimal) measure preserving transformation (and homeomorphism) is said to have quasi-discrete spectrum if it has sufficiently many representations in $\mathcal{C}$. Such transformations have faithful representations in $\overline{\mathcal{C}}$. The class $\overline{\mathcal{C}}$ enjoys A, B, C, D. The "known" maps referred to in B are affine.

This point of view enables us to see the relationship between quasi-discrete spectra and nilflows and unipotent affines on nilmanifolds. Hahn and I [8] showed that, except in trivial cases, nilflow elements could not have quasi-discrete spectrum as the latter transformations cannot be embedded in flows. As we shall point out, nilflows and nilmanifold unipotent affines should be viewed as models generalizing the models defining quasi-discrete spectra.

§5. Nilmanifolds.

Let N be a connected nilpotent Lie group and let $D \subset N$ be a discrete subgroup such that N/D is compact. If A is an automorphism of N preserving D and if $a \in N$ then

$T(x\,D) = a\,A(x)D$ defines an affine transformation of the nilmanifold N/D.

If dA_e is unipotent then T will be called unipotent. If $a_t \in N$ is a one-parameter subgroup of N, then $T_t(x\,D) = a_t\,x\,D$ is called a nilflow. There is a normalized translation invariant measure m on N/D and this measure is preserved by affines.

Let $\mathcal{C}$ be either the class of ergodic unipotent affines or of ergodic nilflows on nilmanifolds. Then $\mathcal{C}$ satisfies A, B, C, in both the topological and measure theoretic contexts. The class $\overline{\mathcal{C}}$ enjoys A, B, D and probably enjoys C but this I am unclear about. The known maps of B are affine in the first case and projective limits of affines in the second. The details are to be found in [9], as is the general procedure for representation as projective limits once condition A is satisfied.

§6. Distal minimal homeomorphisms

In conclusion I should point out that the examples I have

referred to which yield most when subjected to this kind of class analysis are distal minimal homeomorphisms. This is not surprising when we note that the class $\mathcal{C}$ of distal minimal homeomorphisms itself, satisfies A, C, D. $\mathcal{C}$ is probably to large to satisfy B but small enough to make Ellis' proof of C [2] interesting. The proofs of A. and D. are straightforward.

References

1. L. M. Abramov. Metric automorphisms with quasi-discrete spectrum. Izv. Akad. Nauk. SSSR 26(1962) 513-530 = A.M.S. Transl. Ser. 2. 39(1964) 37-56.

2. R. Ellis. Lectures on topological dynamics, Benjamin, New York, 1969.

3. F. Hahn and W. Parry. Minimal dynamical systems with quasi-discrete spectrum. J.L.M.S. 40(1965) 309-323.

4. F. Hahn and W. Parry. Some characteristic properties of dynamical systems with quasi-discrete spectra. Math. Systems. Th.2. (1968) 179-190.

5. P. R. Halmos and J. Von Neumann. Operator methods in classical mechanics. II. Ann. of Math. (43) 1942. 332-350.

6. G. W. Mackey. Ergodic transformation groups with a pure point spectrum. Ill. J. Math. 8(1964) 593-600.

7. D. Ornstein. Factors of Bernoulli shifts are Bernoulli shifts. Advances in Math. (1970) 349-364.

8. W. Parry. Compact abelian group extensions of discrete dynamical systems. Z. Wahrscheinlichkeitstheorie verw. Geb. 13(1969) 95-113.

9. W. Parry. Dynamical representations in nimanifolds. (To appear).

10. R. F. Williams. Classification of symbol spaces of finite type. (To appear.)

SPECTRA OF INDUCED TRANSFORMATIONS

by

Karl Petersen

Let $(X,\mathcal{B},\mu)$ be a probability space, $T: X \to X$ a measure-preserving transformation, and $A \in \mathcal{B}$ a set with positive measure. The derivative transformation $T_A: A \to A$ of T with respect to A is defined as follows. If $A_n = A \cap T^{-1}(X - A) \cap \cdots \cap T^{-n+1}(X - A) \cap T^{-n}A$ for $n = 1,2,\ldots$, then $T_A|A_n = T^n$. Let A' be a copy of A, $\iota: A \to A'$ a one-to-one onto map, and $X^A = X \cup A'$. Then the primitive transformation $T^A: X^A \to X^A$ is defined by $T^A|(X - A) = T|(X - A)$, $T^A|A = \iota$, and $T^A|A' = T\circ\iota^{-1}$. Both T_A and T^A are called induced transformations. When A and X^A are made into probability spaces in the obvious way, then T_A and T^A become measure-preserving transformations which are ergodic if and only if T is ergodic. In case X is a compact Hausdorff space, T is a homeomorphism, and A is open and closed, then T_A and T^A are homeomorphisms of A and X^A, respectively, when these latter spaces are equipped with the obvious topologies. Moreover, the flows (A,T_A) and (X^A,T^A) are minimal (or uniquely ergodic) if and only if the flow (X,T) is minimal (or uniquely ergodic). Detailed information concerning induced transformations and induced flows can be found in [4] and [8].

Kakutani [5] has given examples of induced transformations which are measure-theoretically weakly mixing but not measure-theoretically strongly mixing. The first example is a primitive of the shift transformation on the orbit closure of a certain Toeplitz sequence, and the second is a primitive of the shift on the orbit closure of a sequence similar to the Sturmian sequences studied by Hedlund. These examples are uniquely ergodic minimal flows, and hence they are also topologically weakly mixing; the topological aspects of these and related examples were investigated in [8]. In this note we consider some questions related to the possible existence of measurable or continuous eigenfunctions of induced transformations and induced flows.

1. Almost every derivative transformation is weakly mixing. Let $(X,\mathcal{B},\mu)$ be a Lebesgue space and G the group of (equivalence classes of) measure-preserving transformations $T: X \to X$. For $A \in \mathcal{B}$ with $\mu(A) > 0$, denote by G_A the group of (equivalence classes of) measure-preserving transformations of A. The measure algebra of $(X,\mathcal{B},\mu)$ is a complete metric space under the metric $d(A,B) = \mu(A \Delta B)$, and the group G can be endowed with either of two well-known topologies, the weak topology or the strong topology.

Friedman and Ornstein [1, Theorem 9.3] have proved that if $T \in G$ is ergodic, then the collection of all those $A \in \mathcal{B}$ for which T_A is strongly mixing forms a dense set in the measure algebra of $(X,\mathcal{B},\mu)$. According to a result of Keane [6], if $A \in \mathcal{B}$ then the map $\delta_A: G \to G_A$ defined by $\delta_A(T) = T_A$ is continuous if both G and G_A are given either their strong or weak topologies. A straightforward computation shows that for either choice of the topologies on G and G_A, the inverse image under δ_A of a dense set is dense. It then follows immediately that for a fixed $A \in \mathcal{B}$ with $\mu(A) > 0$, in the weak topology on G the set of all those $T \in G$ for which T_A is weakly mixing is residual, while the set of all those $T \in G$ for which T_A is strongly mixing is first category.

2. Measure-theoretic weak mixing without topological strong mixing. Let $S = \{0,1\}^Z$ be the compact metric space of all bilateral sequences on the symbols 0 and 1, and let $\sigma: S \to S$ be the shift transformation. A sequence $x \in S$ is defined by letting $x(n)$ be 0 if and only if $n = k \cdot 2^{2m} - 1$ for some odd k and some $m = 0,1,2,\ldots$; x is the Toeplitz sequence considered earlier by Kakutani. Let X denote the closure of the orbit of x under σ, and let $A = \{y \in X: y(0) = 0\}$. It is known that the flow (X,σ) is minimal and uniquely ergodic, and Kakutani [5] proved that (X^{X-A},σ^{X-A}) (which is isomorphic to the orbit closure of x with "doubled ones") is measure-theoretically weakly mixing but not measure-theoretically strongly mixing, with respect to its unique invariant measure.

An easy adaptation of Kakutani's proof shows that (A,σ_A) (which amounts to "shifting to the next zero" the sequences in A) is also minimal, uniquely ergodic, and measure-theoretically weakly mixing. This example is of interest because Petersen and Shapiro [8, Theorem 5.1] have proved that the flow (A,σ_A) is not topologically strongly mixing (and hence, incidentally, also not measure-theoretically strongly mixing). It remains to find an example of a uniquely ergodic topologically strongly mixing minimal flow which is not measure-theoretically weakly mixing.

3. Generalized Sturmian sequences. Let $\alpha,\beta \in [0,1)$ with α irrational, and define $x \in S$ by $x(n) = \chi_{[0,\beta)}\langle n\alpha\rangle$, where $\langle a\rangle$ denotes the fractional part of a. The orbit closure X of x under σ was a second example proved by Kakutani to have a primitive transformation which is measure-theoretically weakly mixing but not measure-theoretically strongly mixing, for certain choices of α and β. More generally, let $f(x)$ be a polynomial of positive degree having irrational leading coefficient, and suppose that $\langle f(n)\rangle$ omits the values 0 and β for $n \in Z$.

Let $y(n) = \chi_{[0,\beta)}\langle f(n)\rangle$ for $n \in Z$, $Y = \mathcal{O}(y)^-$, and $A = \{w \in Y: w(0) = 0\}$. Then both (Y^A, σ^A) and (A, σ_A) are minimal, uniquely ergodic, and topologically weakly mixing. This is proved by considering almost-automorphic extensions of a certain affine transformation on an r-torus, of the type studied earlier by F. Hahn [3, §5].

4. Measurable eigenfunctions of topologically weakly mixing flows. As in §3 above, let $\alpha, \beta \in [0,1)$, suppose α is irrational, define $x \in S$ by $x(n) = \chi_{[0,\beta)}\langle n\alpha\rangle$, and let $X = \mathcal{O}(x)^-$ and $A = \{y \in X: y(0) = 0\}$. Petersen and Shapiro [8] showed that if $\beta \notin Z\alpha$, then (X^{X-A}, σ^{X-A}) is minimal and topologically weakly mixing. Indeed, Furstenberg, Keynes, and Shapiro [2] proved that if also $2\beta \notin Z\alpha$, then (X^{X-A}, σ^{X-A}) is a prime flow, in the sense that it has no proper homomorphic images.

We observe that it follows already from a result of Veech [9, Theorem 3] that, for certain choices of α and β, (X^{X-A}, σ^{X-A}) has a measurable eigenfunction with eigenvalue -1; thereby we obtain an example of a topologically weakly mixing (even prime) flow which is not measure-theoretically weakly mixing.

Formal manipulation of Fourier series as in [7] shows that if α and β can be chosen so that

$$\theta(x) = \frac{1}{1+\beta} \sum_{n \neq 0} \frac{1}{2\pi in} \frac{1 - e^{-2\pi in\beta}}{e^{2\pi in\alpha} - 1} e^{2\pi inx}$$

belongs to $L^2[0,1)$, then (X^{X-A}, σ^{X-A}) has a measurable eigenfunction $f(x) = e^{2\pi i \operatorname{Re} \theta(x)}$ with eigenvalue $e^{2\pi i/(1+\beta)}$. Consequently, if β can also be chosen to be irrational, then (X^{X-A}, σ^{X-A}) has countably many measurable eigenfunctions but no continuous eigenfunctions. We do not know the answer to the following interesting question of Veech: can α and β be chosen so that $2\beta \notin Z\alpha$ and (X^{X-A}, σ^{X-A}) has discrete spectrum?

REFERENCES

1. N. Friedman, Introduction to Ergodic Theory, Van Nostrand Reinhold Co., New York, 1970.

2. H. Furstenberg, H. B. Keynes, and L. Shapiro, Prime flows in topological dynamics, to appear.

3. F. Hahn, On affine transformations of compact abelian groups, Amer. J. Math. 85 (1963), 428-446.

4. S. Kakutani, Induced measure preserving transformations, Proc. Imp. Acad. Tokyo 19 (1943), 635-641.

5. S. Kakutani, Weakly mixing dynamical systems which are not strongly mixing, seminar talk, Yale University, 1968.

6. M. Keane, Contractibility of the automorphism group of a non-atomic measure space, Proc. Amer. Math. Soc. 26 (1970), 420-422.

7. A. N. Kolmogorov, On dynamical systems with an integral invariant on the torus, Doklady Akad. Nauk. U.S.S.R. 93 (1953), 763-766. (Russian).

8. K. Petersen and L. Shapiro, Induced flows, to appear in Trans. Amer. Math. Soc.

9. W. A. Veech, Strict ergodicity in zero dimensional dynamical systems and the Kronecker-Weyl theorem mod 2, Trans. Amer. Math. Soc. 140 (1969), 1-33.

UNIVERSITY OF NORTH CAROLINA
CHAPEL HILL, NORTH CAROLINA 27514

ASYMPTOTIC CYCLES FOR DISCRETE FLOWS

F. Rhodes

University of Southampton

1. Introduction

In [5], Schwartzman defined asymptotic cycles for continuous flows on compact metric spaces. In [4], asymptotic cycles have been defined for certain continuous curves on geodesic spaces, whether or not they arise as orbits of a continuous flow. A natural approach to asymptotic cycles for discrete flows is to suspend a discrete flow (X,Z) in a continuous flow (Y,R). However, the homology of the space Y then depends on the action of Z on X as well as on the homology of X, and there is in general no natural way of lifting back an asymptotic cycle from $H_1(Y,R)$ to $H_1(X,R)$.

An alternative approach is to associate with an orbit of a discrete flow a family of continuous curves, and to study the resulting family of asymptotic cycles. This is the approach which is outlined in this note. The full results will be published elsewhere.

2. μ-recurrent flows

The fundamental group $\sigma(X,x_0,G)$ of a transformation group (X,G) was introduced in [2]. For a discrete flow (X,Z) generated by a homeomorphism ϕ the elements of $\sigma(X,x_0,Z)$ are homotopy classes $[f;n]$ where f is a path from x_0 to $\phi^n x_0$. The rule of composition is

$$[f;n] * [g;m] = [f + \phi^n g ; m + n].$$

A path f from x_0 to $\phi^n x_0$ gives rise to an infinite curve α which can be thought of intuitively as $f + \phi^n f + \phi^{2n} f + \dots$. Precisely,

$$\alpha : [0,\infty) \to X,$$

$$\alpha : \lambda \to \phi^{n|\lambda|}(\lambda - |\lambda|),$$

where $|\lambda|$ denotes the integer part of λ.

One of the properties of a geodesic space (i.e. a G-space in the sense of Busemann [1]) is that for each point X there is a number $\rho(x) > 0$ such that each pair of points in the spherical neighborhood $S(x,\rho(x))$ is joined by a unique segment. This neighborhood is simply-connected, so a path from x_o which ends at a point of $S(x_o,\rho(x_o))$ gives rise to an element of $H_1(X,R)$. If a curve has the property that infinitely often each point returns within a simply-connected neighborhood of itself, then one can establish close connections between the homology elements which arise in this way. The following definition will be required.

<u>Definition 1</u>. A homeomorphism ϕ of a metric space X will be said to be <u>μ-recurrent</u> if the set

$$E(\phi,\mu) = \{n \in N \mid \forall x \in X, \phi^n x \in S(x,\mu)\}$$

contains a sequence $n_i \to \infty$.

The positive semi-orbit of a point x_o under a homeomorphism ϕ will be said to be <u>μ-recurrent</u> if the set

$$E(\phi,x_o,\mu) = \{n \in N \mid \forall r \in N, \phi^{n+r}x_o \in S(\phi^r x_o,\mu)\}$$

contains a sequence $n_i \to \infty$.

Suppose now that $\inf \rho(x) > \mu > 0$ and that ϕ is μ-recurrent. Then ϕ^{-1} is also μ-recurrent. If f is a path from x_o to $\phi^n x_o$ and α is the corresponding infinite curve, then for each $t \in E(\phi,\mu)$ the path $\alpha[0,t]$ followed by the segments from $\phi^{it}x_o$ to $\phi^{(i-1)t}x_o$, $i = n, n-1, \ldots, 2, 1$, is a cycle whose homology class (with real coefficients) will be denoted by $v(f;n:t)$. Moreover, if $s,t \in E(\phi,\mu)$ and k is an integer such that $ks \leq t < (k+1)s$, then it can be shown that $v(f;n:t) = kv(f;n:s) + w$, where w is the homology class of a loop of length less than $n(k+1)\mu + \ell(s)$ and $\ell(s)$ is independent of t. For $n = 1$ the proof of this statement is contained in the proof of Prop. 3.1 and the introductory remarks of §4 of [4].

To ensure that $v(f;n:t)/t$ tends to a limit as $t \to \infty$ through values in $E(\phi,\mu)$ we require the space to satisfy an extra condition. It serves to guarantee that a short cycle gives rise to an element of the vector space $H_1(X,R)$ whose ℓ^1 norm (relative to any basis for the first Betti group) is small. Let Γ_m denote the set of rectifiable cycles whose homology class has norm m (relative to a given basis). Let $\ell(f)$ denote the length of f, and let $L_m = \inf \ell(f)$, $f \in \Gamma_m$. The number $L = \inf L_m/m$, $m > 0$ is called the <u>mean cycle length</u> (relative to the given basis). If $H_1(X,R)$ is finitely generated, the condition $L > 0$ is independent of the chosen basis. Moreover, if $L > 0$ then the norm of the homology class of a rectifiable cycle f is not greater than $\ell(f)/L$.

We now have the following theorem.

<u>Theorem 1</u>. Let X be a geodesic space for which $\inf \rho(x) > \mu > 0$. Let $H_1(X,R)$ be finitely generated, and let the mean cycle length of X be positive. Let ϕ be a μ-recurrent homeomorphism of X. Then as $t \to \infty$ through values $t \in E(\phi,\mu)$, $v(f;n:t)/t$ tends to a limit $v[f;n]$ which depends only on the class $[f;n]$. If $[f;n]$ commutes with $[g;m]$ in $\sigma(X,x_o,Z)$, then

$$v([f;n] * [g;m]) = v[f;n] + v[g;m].$$

It follows that if $\sigma(X,x_o,Z)$ is abelian then the map

$$v : \sigma(X,x_o,Z) \to H_1(X,R)$$

$$v : [f;n] \to v[f;n]$$

is a homomorphism. However, the condition is quite restrictive. The group $\sigma(X,x_o,Z)$ is abelian if and only if $\pi_1(X,x_o)$ is abelian and ϕ induces the identity map on $H_1(X)$. In this case $\sigma(X,x_o,Z)$ is isomorphic to $\pi_1(X,x_o) \times Z$. Moreover, as is proved in [3], $\sigma(X,x_o,Z)$ is isomorphic to the fundamental group of the suspension space Y. Thus when $\sigma(X,x_o,Z)$ is abelian one could get some information via the technique of suspending the discrete flow in a continuous flow.

3. <u>μ-recurrent orbits</u>

We now turn to the case in which ϕ need not be μ-recurrent, but there is a point x_0 whose semi-orbit is μ-recurrent. It will be necessary to compensate for the weakening of the recurrence condition by imposing an equicontinuity type condition. One form of the result is stated here.

Definition 2. The maps $\{\phi^n | n \in N\}$ will be said to be (ν,μ)-equicontinuous if $d(x,y) < \nu$ implies $d(\phi^n x, \phi^n y) < \mu$, $n \in N$.

Let $H_1(X,R)$ be finitely generated and let the mean cycle length of X be positive. Then a homeomorphism ϕ whose iterates are (ν,μ)-equicontinuous, for some numbers ν and μ, induces a periodic automorphism ϕ_* of $H_1(X,R)$.

Theorem 2. Let X be a geodesic space for which $\inf \rho(x) > \mu > 0$. Let $H_1(X,R)$ be finitely generated and let the mean cycle length of X be positive. Let the maps $\{\phi^n | n \in N\}$ be (ν,μ)-equicontinuous and let the positive semi-orbit of x_0 be ν-recurrent. If $s,t \in E(\phi,x_0,\nu)$ and $ks \leq t < (k+1)s$, then for each element $[f;n]$ for which n is positive

$$v(f;n:t) = (1 + \phi_*^{ns} + \dots + \phi_*^{n(k-1)s}) v(f;n:s) + w$$

where the norm of the element w/t of $H_1(X,R)$ can be made arbitrarily small by choice of s and t. If the period of ϕ_* is p and p divides n then $v(f;n:t)/t$ tends to a limit $v[f;n]$ as $t \to \infty$ through values $t \in E(\phi,x_0,\nu)$.

REFERENCES

1. Herbert Busemann, The Geometry of Geodesics, Academic Press, New York, 1955.

2. F. Rhodes, On the Fundamental Group of a Transformation Group, Proc. London Math. Soc. (3) 16 (1966) 635-50.

3. F. Rhodes, The Mapping Torus of a Transformation Group, Quart. J. Math. Oxford (2) 20 (1969) 45-58.

4. F. Rhodes, Asymptotic Cycles for Continuous Curves on Geodesic Spaces, J. London Math. Soc. (to appear).

5. Sol Schwartzman, Asymptotic Cycles, Ann. Math. (2) 66 (1957) 270-284.

ABELIAN SEMI-GROUPS OF EXPANDING MAPS

RICHARD SACKSTEDER

1. Introduction. A great deal of recent work has been concerned with the problem of discovering large classes of maps that are topologically stable and can be classified up to topological conjugacy (see [1], [7], or [10]). It is necessary that the problem be posed in terms of topological stability and conjugacy, since there seem to be no interesting classes of maps that are differentiably stable, and only a few [redacted] (cf. [2], [6], [12], [13]) that can be classfied up to differentiable conjugacy in a satisfactory way.

The purpose of this paper is to show that the situation changes radically if one considers suitable semi-groups of maps rather than a single map and its iterates. Our concern here will be essentially with expanding maps of $S^1 \approx R^1/Z$. Recall that a C^1 map, t, is an expanding map if

$$\limsup \|Dt^n(X)\|^{1/n} / \|X\|^{1/n} > \lambda \quad \text{as} \quad n \to \infty$$

holds for every tangent vector X, where λ is a constant greater than one, and $\| \ \|$ denotes length in some Riemannian metric. It is well-known that it is always possible to find a metric such that

$$\|Dt(X)\| \geq \lambda \ \|X\|$$

holds for every X, when t is defined on a compact manifold.

Our starting point is the following very special case of a theorem of Shub [10].

Proposition 1.1: *Let t be an expanding map of S^1. Then there is a homeomorphism h of S^1 satisfying $h(t(x)) = Nh(x)$, (N = degree t).*

(Here and below the equality is to be understood mod 1.) Hirsch has observed [4, p.125] that it is not possible to strengthen the conclusion to make h of class C^1, even if t is analytic. The theorems proved here are stated in terms of the class C^{n-} of smoothness, by which is meant the class of functions having $n-1$ Lipschitz continuous derivatives. The following theorem is our main result.

Theorem 1.2: Let t_N and t_M be expanding maps of S^1 that commute, have degrees N and M respectively, and are of class C^{n-} $(n \geqq 2)$. Then if N and M are relatively prime, there is a diffeomorphism h of S^1 of class C^{n-} satisfying $h(t_N(x)) = Nh(x)$ and $h(t_M(x)) = Mh(x)$.

The whole point of the theorem is that, in contrast to the case of the proposition, it can be asserted that h is smooth. The theorem has the following consequence.

Corollary 1.3: Let A be an Abelian semi-group of expanding maps. Suppose that A contains elements t_N and t_M satisfying the conditions of Theorem 1.2, and let h be as in that theorem. Then every element t of A satisfies $h(t(x)) = Kh(x)$, where K = degree t.

This corollary obviously admits an interpretation as a stability theorem for representations of Abelian semi-groups in addition to its interpretation as a conjugacy theorem.

A slightly stronger version of Theorem 1.2 is proved in Section 4. Most of the remainder of the paper is concerned with infinitesimal stability. The main result in this direction is Theorem 5.1, which suggests that it should be possible to generalize Theorem 1.2 further.

There seems to have been very little previous work on Abelian semi-groups of maps in the spirit of this paper, although [3] and [5] are somewhat related. The results obtained here for S^1 suggest that the higher dimensional cases are worth investigating.

2. Lemmas. The following lemma is a special case of a theorem proved in [8]. (cf. also Theorem 3.1 of [9].)

Lemma 2.1: Let t be an expanding map of S^1 of class C^{n-} ($n \geqq 2$). Then there is a unique measure m that is: (i) invariant under t, (ii) such that $m(S^1) = 1$, and (iii) absolutely continuous with respect to Lebesgue measure. Moreover the distribution function of is of class C^{n-}.

Here by the distribution function is meant the function H defined in the fundamental domain $[0,1)$ of S^1 by $H(x) = m([0,x))$. By "m is invariant under t" is meant that the measure $t_* m$ defined by $(t_* m)(E) = m(t^{-1}(E))$ is the same as m.

Lemma 2.2: Let t be as in Lemma 2.1. Then since $H(0) = 0$ and $H(1) = 1$, H defines a diffeomorphism h of S^1 of class C^{n-}, and $h_* m$ is Lebesgue measure, hence hth^{-1} leaves Lebesgue measure invariant.

Proof: It is convenient to let the point 0 of $[0,1)$ correspond to a fixed point of t. It is to be shown that $h_* m([0,x)) = x$ for any x, $0 \leqq x < 1$. But $h_* m([0,x)) = m(h^{-1}([0,x))) = m([0,H^{-1}(x))) = H(H^{-1}(x)) = x$.

Lemma 2.3: Let t and h be as in Lemmas 2.1 and 2.2. Suppose that s is any onto map of class C^1 with non-vanishing Jacobian that commutes with t. Then hsh^{-1} leaves Lebesgue measure invariant.

Proof: The conclusion is equivalent to $s_* m = m$, but the latter is a consequence of the uniqueness part of Lemma 2.1 as follows: $t_* s_* m = s_* t_* m = s_* m$, so $s_* m$ is invariant under t. The conditions on s assure that $s_* m$ is absolutely continuous with respect to Lebesgue measure, hence $s_* m = m$ follows by uniqueness.

<u>Lemma 2.4: Let</u> t <u>be an expanding map of</u> S^1 <u>that satisfies</u> $t(Nx)) = Nt(x)$ <u>identically in</u> x <u>for some integer</u> N <u>with</u> $|N| > 1$. <u>Then</u> $t(x) = Mx$, <u>where</u> $M =$ <u>degree</u> t .

We omit the proof of Lemma 2.4, since it is essentially the same as the proof of Theorem 2 of [10] (cf. [7], p.65), specialized to S^1 .

3. <u>Proof of Theorem 1.2.</u> Let h be the map defined in Lemma 2.2, and define $w_N = ht_N h^{-1}$ and $w_M = ht_M h^{-1}$. Then Lemmas 2.2 and 2.3 show that w_N and w_M satisfy the hypotheses of Theorem 1.2 and, in addition, leave Lebesgue measure invariant. The theorem will be proved by showing that $w_M(x) = Mx$, hence by Lemma 2.4, $w_N(x) = Nx$.

Let W_N and W_M be the lifts of w_N and w_M to R^1 . It can be supposed that $W_N(0) = W_M(0) = 0$. It will also be supposed that $N > 1$, since the case $N < -1$ requires slight modifications in notation. Let $D_M = dW_M/dx$ and $a_i = W_N^{-1}(i)$ for $i = 0,1,\ldots,N$. Clearly, $a_0 = 0 < a_1 < \ldots < a_N = 1$. Let I_i be the image of $[a_{i-1}, a_i)$ in S^1 . Then the restriction of w_N to I_i is one-one onto $[0,1)$ (viewed as a fundamental domain for S^1), and $S^1 = \bigcup \{I_i : i = 1,\ldots,N\}$. Let x_i denote the inverse function to $w_N | I_i$, which is also of class C^{n-} .

Let V_N denote the isometry of $L^2(S^1)$ defined by $V_N f(x) = f(w_N(x))$. Then the adjoint V_N^* of V_N is defined by the requirement that for every f and g in $L^2(S^1)$,

$$(3.1) \qquad \int_0^1 V_N f(x)g(x)dx = \int_0^1 f(x)V_N^* g(x)dx .$$

V_N^* is described more explicitly by the formula

$$(3.2) \qquad V_N^* g(x) = \sum_{i=1}^{n} g(x_i(x))x_i'(x) ,$$

which follows from (3.1) by taking f to be the characteristic function of the interval $[0,y)$ and differentiating both sides with respect to y.

Taking f to be the characteristic function of $[0,y)$ and taking $g \equiv 1$ in (3.1) gives

$$y = \text{measure}(w_N^{-1}(0,y)) = \int_0^y V_N^* 1(x)dx ,$$

where the first equality comes from the fact that w_N preserves Lebesgue measure. Differentiating both sides gives

$$V_N^* 1 \equiv 1 . \tag{3.3}$$

The main part of the proof consists in showing that

$$V_N^* D_M = D_M . \tag{3.4}$$

By (3.2), the left side of (3.4) is equal to

$$\sum_{i=1}^{N} D_M(x_i(x))x_i'(x) = d/dx \sum_{i=1}^{N} W_M(x_i(x)) .$$

For any i, there exists a $j = j(i)$ such that $w_M(x_i(x)) = x_j(w_M(x))$. This is the case because $\{x_j(w_M(x)): j = 1,\dots,N\}$ is just the inverse image of $w_M(x)$ under w_N, and $w_N w_M(x_i(x)) = w_M w_N(x_i(x)) = w_M(x)$. Moreover, we shall show that the correspondence between i and $j(i)$ is one-one. Since this is a purely topological assertion, it can be assumed temporarily that $w_N = Nx$ and $w_M = Mx$ by Proposition 1.1 and Lemma 2.4. Then $w_M(x_{i+1}(x)) = M(x+i)/N$ and $x_{j+1}(w_M(x)) = (Mx+j)/N$. Now it is clear that the assertion follows from the fact that when M and N are relatively prime, multiplication by M permutes the residue classes modulo N. Therefore we can now write

$$V_N^* D_M(x) = d/dx \sum_{i=1}^{N} x_i(w_M(x)) = D_M(x) \sum_{i=1}^{N} x_i'(w_M(x)) \quad .$$

But (3.2) and (3.3) imply that the right side is just $D_M(x)$. This proves (3.4).

To complete the proof observe first that iterating (3.4) gives $D_M(x) = (V_N^*)^k D_M(x)$ for $k = 1,2,\ldots,$ while Theorem 4.1 of [9] shows that

$$\lim_{k\to\infty} (V_N^*)^k D_M(x) = \int_0^1 D_M(y)dy = M \quad .$$

Therefore $D_M(x) \equiv M$ and the theorem is proved.

4. <u>An Extension of Theorem 1.2.</u> Let $\{N_1, N_2, \ldots\}$ be a set of integers, each of absolute value greater than one. By $A(N_1, N_2, \ldots)$ will be meant the smallest multiplicative sub-semi-group, A , of the non-zero integers satisfying the condition: If N and M are in A , M divides N , and $|N/M| > 1$, then N/M is in A . Two examples will illustrate the concept. $A(6,10,15) = \{6^a 10^b 15^c 4^d 9^e 25^f\}$ and $A(6\times 7, 10\times 11, 15\times 13) = \{(6\times 7)^a (10\times 11)^b (15\times 13)^c\}$. Of course, all of the exponents are to be non-negative and not all are allowed to be zero simultaneously. In both cases, the numbers defining A have greatest common divisor one and no pair of them is relatively prime. However, in the first case, but not in the second, A contains a relatively prime pair of integers.

To understand the relevance of $A(N_1, N_2, \ldots)$ for our purposes, first note that Shub's Theorem (Proposition 1.1 above) and Lemma 2.4 imply that any commuting family of expanding maps of S^1 can be imbedded in a commuting family $\mathcal{J}$ of covering maps containing exactly one element of degree N for every $N \neq 0$, ± 1 . If integers $N_1, N_2, \ldots$ correspond to maps in

$\mathcal{T}$ which are of class $C^{n-}(n \geq 2)$, then the maps in $\mathcal{T}$ of degree N for any N in $A(N_1,N_2,\ldots)$ are of class C^{n-}. This assertion is clear, because: (1) any element of a semi-group generated by a set of C^{n-} maps is itself C^{n-}, and (2) if t_N and t_M are C^{n-} maps of degree N and M , and M divides N (say $N/M = K$), then $t_N = t_M t_K$ where t_K is of degree K . But t_M has local inverses of class C^{n-}, so t_K is of class C^{n-}.

The above remarks together with Corollary 1.3 prove the following:

Corollary 4.1: Let $t_1,t_2,\ldots$ be C^{n-} $(n \geq 2)$ expanding maps of S^1 that commute pairwise. Let $N_1,N_2,\ldots$ be the degrees of $t_1,t_2,\ldots$ respectively. Then if $A(N_1,N_2,\ldots)$ contains a pair of relatively prime integers, there is a C^{n-} diffeomorphism h of S^1 such that $h(t_j(x)) \equiv N_j h(x)$ for $j = 1,2,\ldots$.

5. Infinitesimal Stability. The concepts of conjugacy and stability have infinitesimal analogues, which will now be defined for maps of S^1 . Suppose that $\mathcal{T} = \{t_1,\ldots,t_k\}$ is a family of C^{n-} maps of S^1 , where $n \geq 2$. Let $D_i = \partial t_i/\partial x$. A set $\mathcal{P} = \{G_1,\ldots,G_k\}$ of k real valued functions of class C^{m-} defined on S^1 will be called a C^{m-} infinitesimal perturbation of $\mathcal{T}$ provided for every i and j with $1 \leq i < j \leq k$,

$$(5.1) \qquad G_i(t_j(x)) - D_j(t_i(x))G_i(x) = G_j(t_i(x)) - D_i(t_j(x))G_j(x) .$$

The motivation for this concept is as follows: Suppose that for $|s| < \varepsilon$, a family $\mathcal{T}^s = \{t_i^s,\ldots,t_k^s\}$ of maps of S^1 whose first derivative with respect to s exists and is of class C^{m-} in x is given. If $\mathcal{T}^0 = \mathcal{T}$, and for each s the maps commute pairwise, (5.1) results from differentiating $t_i^s(t_j^s(x)) = t_j^s(t_i^s(x))$ with respect to s at $s = 0$, if $G_i = \partial t_i^s/\partial s$ at $s = 0$.

An infinitesimal perturbation $\mathcal{P}$ of $\mathcal{J}$ is said to be an infinitesimal C^{n-} conjugacy if there is a C^{n-} function F defined on S^1 such that for $i = 1,\dots,k$,

$$F(t_i(x)) - D_i(x)F(x) = G_i(x). \tag{5.2}$$

The motivation for this definition is that when a family $\mathcal{J}^s$ is obtained from $\mathcal{J}$ by setting $t_i^s(x) = h_s(t_i(h_s^{-1}(x)))$, where h_s is a family of diffeomorphisms of S^1 with $h_0(x) \equiv x$, differentiating $t_i^s(x)$ with respect to s at $s = 0$ gives (5.2), where $F(x) = \partial h/\partial s(x)$ at $s = 0$.

Finally, a family $\mathcal{J}$ as above is called (C^{n-}, C^{m-}) infinitesimally stable if every C^{m-} infinitesimal perturbation is a C^{n-} infinitesimal conjugacy.

The following theorem will be proved in the next two sections.

Theorem 5.1: Let $t_i(x) = N_i x$, where $|N_i| > 1$, for $i = 1,\dots,k$. Suppose that one is the greatest common divisor of $|N_1|,\dots,|N_k|$. Then the family $\mathcal{J} = \{t_1,\dots,t_k\}$ is (C^{n-}, C^{n-}) infinitesimally stable for $n \geqq 2$.

6. Solutions of Functional Equations. In this section M and N will denote integers of absolute value greater than one. if $w_N(x) = Nx \pmod 1$, the isometry V_N defined in Section 3 becomes $(V_N f)(x) = f(Nx)$ and its adjoint is given by $(V_N^* f)(x) = N^{-1} \sum_{i=1}^{N} f((x+i)N^{-1})$.

Lemma 6.1: Let G_N: S^1 , R^1 be continuous. Then

(6.1) $$V_N F_N - NF_N = G_N$$

has a unique continuous solution, F_N . F_N need not be absolutely continuous even if G_N is analytic.

Proof: Iteration of (6.1) gives

$$F_N(x) = -\sum_{j=0}^{k-1} N^{-j-1} G_N(N^j x) + N^{-k} \quad F_N(N^k x) \quad .$$

Since $|N| > 1$, $F_N(x) = -\sum_{j=0}^{\infty} N^{-j-1} G_N(N^j x)$ is the unique solution of (6.1).

Let $e(x) = \exp(2\pi i x)$. If $G_N(x) \sim \Sigma\, a_j e(jx)$ and $F_N(x) \sim \Sigma\, b_j e(jx)$, the coefficients are related by $b_0 = (1 - N)^{-1} a_0$ and

(6.2) $$b_n = -\sum_{j=0}^{k} N^{-j-1} a_{n/N^j} \quad ,$$

where k is defined by $N^k | n$, (i.e., N^k divides n) and $N^{k+1} \nmid n$.

Now suppose that G_N is defined by taking for some constant θ , $0 < \theta < 1$, $a_n = \theta^{i-1}$ when $|n| = |N|^i$ and $a_n = 0$ otherwise. Then G_N is obviously analytic, but (6.2) shows that if $|n| = |N|^r$,

$$|nb_n| = |N|^r \sum_{j=0}^{r} |N|^{-r-1+j} \theta^{j-1} = \sum_{j=0}^{r} |N\theta|^{j-1} \quad .$$

Therefore nb_n does not approach zero, so F_N is not absolutely continuous.

Remark: The lemma just proved is the infinitesimal analogue of Proposition 1.1 together with Hirsch's observation about the lack of smoothness of Shub's map.

If F_N and G_N have derivatives f_N and g_N , (6.1) implies that $V_N f_N - f_N = N^{-1} g_N$. Applying V_N^* to both sides gives

$(6.3)_N$ $$f_N - V_N^* f_N = N^{-1} V_N^* g_N \ , \text{ where } \int_0^1 f_N(x)dx = 0 \ .$$

The following lemma about the solutions of $(6.3)_N$ is proved in [8] .

Lemma 6.2: Let g_N be of class C^{n-} $(n \geqq 1)$ and satisfy

$(6.4)_N$ $$\int_0^1 g_N(x)dx = 0 \ .$$

Then there is a unique f_N of class C^{n-} satisfying $(6.3)_N$.

Lemma 6.3: Let g_N and g_M be of class C^{n-} $(n \geqq 1)$ and satisfy $(6.4)_N$ and $(6.4)_M$. Suppose that

(6.5) $$M(V_M g_N - g_N) = N(V_N g_M - g_M) \ .$$

Then the solutions f_N and f_M of $(6.3)_N$ and $(6.3)_M$ agree.

Proof: Let $h_N = N(V_N f_N - f_N)$, and $h_M = M(V_M f_N - f_N)$.

Clearly $(6.3)_N$ implies that $V_N^* g_N = V_N^* h_N$. Then it is easy to check that

(6.6) $$N(V_N h_M - h_M) = M(V_M h_N - h_N) \ .$$

Letting $Q_N = h_N - g_N$ and $Q_M = h_M - g_M$, (6.5) and (6.6) imply that

(6.7) $$N(V_N Q_M - Q_M) = M(V_M Q_N - Q_N) \ .$$

Applying $V_N^* V_M^* = V_M^* V_N^*$ to both sides of (6.7) gives

(6.8) $$V_M^* Q_M - V_N^* V_M^* Q_M = 0 \ ,$$

since $V_N^* Q_N = 0$. The equation (6.8) is just the case $f_N = V_M^* Q_M$ and

$g_N = 0$ of $(6.3)_N$. The uniqueness part of Lemma 6.2 therefore implies that $V_M^* Q_M = 0$, that is $V_M^* g_M = V_M^* h_M = M(f_N - V_M^* f_N)$. Now the uniqueness part of Lemma 6.2 applied to $(6.3)_M$ shows that $f_N = f_M$.

Lemma 6.4: Assume the hypotheses of Lemma 6.3. Let $Q_M(x) \sim \Sigma\, a_n e(nx)$. Then $a_n \neq 0$ can only occur if $M \nmid m$, but $M \mid mN^q$ for some $q \geqq 1$.

Proof: Let $Q_N(x) \sim \Sigma\, b_n e(nx)$ and let either side of (6.7) have expansion $\Sigma\, c_n e(nx)$. Then (6.7) gives the relations

(6.9) $$a_n = -N^{-1} c_n + a_{n/N},$$

where the convention is understood that $a_{n/N}$ is taken as zero if $N \nmid n$. Similarly,

(6.10) $$b_n = -M^{-1} c_n + b_{n/M}.$$

Note that since $N^{-1} \sum_{j=1}^{N} e(n(x+j)N^{-1}) = 1$, $V_N^* Q_N = 0$ implies that

(6.11) $$b_n = 0 \text{ if } N \mid n .$$

Similarly,

(6.12) $$a_n = 0 \text{ if } M \mid n .$$

Now the conditions on m can only fail to hold if $M \mid m$ or if $M \nmid mN^q$ for every positive q. In the first case, (6.12) shows that $a_m = 0$. In the second, (6.10) and (6.11) show that for all positive q, $b_{mN^q} = -M^{-1} c_{mN^q} = 0$, hence (6.9) gives

$$a_{mN^q} = a_m - \sum_{j=1}^{q} c_{mN^j} = a_m .$$

But the Riemann-Lebesgue Lemma implies that this is possible only if $a_m = 0$.

7. <u>Proof of Theorem 5.1.</u> Let $\{G_1,\ldots,G_k\}$ be an infinitesimal C^{n-} perturbation of $t_1,\ldots,t_k$. It is to be shown that there is an F of class C^{n-} satisfying

$$(7.1)_i \qquad F(N_i x) - N_i F(x) = G_i(x) , \quad \text{for } i = 1,\ldots,k .$$

Let $f = dF/dx$ and $g_i = dG_i/dx$. Differentiating $(7.1)_i$ gives

$$(7.2)_i \qquad N_i(f(N_i x) - f(x)) = g_i(x) \quad \text{for } i = 1,\ldots,k .$$

Since the G_i's satisfy (5.1), the g_i's satisfy

$$(7.3) \qquad N_i(g_i(N_i x) - g_i(x)) = N_j(g_j(N_j x) - g_j(x)) \quad \text{for } 1 \leqq i < j \leqq k .$$

Lemma 6.3 then shows that there is an f of class $C^{(n-1)-}$ such that $(6.3)_N$ is satisfied by $f = f_N$ for $N = N_1,\ldots,N_k$.

Now we want to show that this f also satisfies $(7.2)_i$ for $i = 1,\ldots,k$. Set $M = N_k$ and apply Lemma 6.4 (k-1) times taking $N = N_1,\ldots,N_{k-1}$. It follows that $a_m \neq 0$ is possible only if $M \nmid m$, but $M | mN_i^q$ for some positive $q = q_i$, where $i = 1,\ldots,k-1$. But this contradicts g.c.d. $(|N_1|,\ldots,|N_{k-1}|,|M|) = 1$. Therefore, $a_m = 0$ for every m, and $Q_M(x) = M(V_M f - f) - g_M = 0$. This is just $(7.2)_k$. Similarly $(7.2)_i$ holds for $i = 1,\ldots,k-1$. Integrating $(7.2)_i$ gives $(7.1)_i$ and the proof is complete.

REFERENCES

[1] American Mathematical Society, Proceedings of Symposia in Pure Mathematics, vol. 14, (1970).

[2] Arnold, V.I., "Small denominators, I, on the mapping of a circle into itself", American Mathematical Society Translations, 2nd Series, vol. 46, pp. 213-284.

[3] Furstenberg, H., "Disjointness in ergodic theory, minimal sets, and a problem in Diophantine approximation", (especially part IV), Mathematical Systems Theory, vol. 1, (1967), pp. 1-49.

[4] Hirsch, M.W., "Expanding maps and transformation groups", In [1], pp. 125-131.

[5] Kopell, N., "Commuting diffeomorphisms", In [1], pp. 165-184.

[6] Moser, J., "A rapidly convergent iteration method and non-linear differential equations - II", Annali Scuola Normale Superiore di Pisa, Series 3, vol. 20, (1966), pp. 499-535.

[7] Nitecki, Z., Differentiable Dynamics, The M.I.T. Press, (1971).

[8] Sacksteder, R., "The measures invariant under an expanding map", to appear.

[9] Sacksteder, R., "On convergence to invariant measures", to appear.

[10] Shub, M., "Endomorphisms of compact differentiable manifolds", The American Journal of Mathematics, vol. 91, (1969), pp.175-199.

[11] Smale, S., "Differentiable dynamical systems", Bulletin of the American Mathematical Society, vol. 73, (1967), pp.747-817.

[12] Sternberg, S., "Celestial Mechanics, Part II, W.A. Benjamin, Inc., New York, (1969).

[13] Tischler, R., "Conjugacy problems for R^k actions", Dissertation, The City University of New York, (1970).

Graduate Center
City University of New York

IRREGULARITIES OF DISTRIBUTION
IN DYNAMICAL SYSTEMS

LEONARD SHAPIRO*

One of the original goals of the theory of dynamical systems was to find a method for determining the "visitation time" of a point x_o within a subset A of the phase space X. That is, given a dynamical system $T:X \to X$, $A \subseteq X$, and $x_o \in X$, what is

$$\lim_{n\to\infty} \frac{1}{n+1} \sum_{i=0}^{n} \chi_A(T^i x_o) \quad ?$$

The ergodic theorem states that under certain conditions this limit is the measure of A, for almost all x_o. Thus if we are given sets A and B of equal measure, almost all points x_o visit A and B with the same frequency. In this paper we are concerned with a more delicate estimate of the difference in visitation times, namely we ask whether

$$(*) \qquad \sum_{i=0}^{n} \chi_A(T^i x_o) - \chi_B(T^i x_o)$$

is bounded for $n \in \mathbf{Z}$. It is clear that if $B = T^k A$, then the absolute value of (*) is always bounded by $|k|$. We shall see that in some cases, e.g. when (X, T) is the irrational rotation on a circle and A, B are intervals, that this is essentially the only case when (*)

* Supported by NSF grant 20871

is bounded.

Our methods are topological and we can obtain information about specific points x_o , as opposed to the measure-theoretic setting where information is usually available only almost everywhere.

We consider a dynamical system (X, T) where T is a homeomorphism of the compact Hausdorff space X , such that (X, T) is minimal, and we fix $x_o \in X$. For fixed subsets A and B of X we define $g(T^n x_o) = \chi_A(T^n x_o) - \chi_B(T^n x_o)$ on the orbit of x_o , and let F denote the set of points x in X to which g can not be extended continuously. Setting $m_o(n) = g(T^n x_o)$ we obtain an element m_o of the space of bisequences on the symbols $\{1, 0, -1\}$. We let M denote the σ-orbit closure of m_o in this space, where σ is the left shift, $\sigma m(n) = m(n+1)$.

Theorem We make the following assumptions:

1) (M, σ) is minimal,

2) The map $\pi(\sigma^n m_o) = T^n x_o$ can be extended to a homomorphism $\pi: (M, \sigma) \to (X, T)$,

3) there is $x_1 \in X$ whose orbit does not intersect F , and

4) there is $z \in X$ with $\mathcal{O}(z) \cap F = \{z\}$.

Then $(*)$ is unbounded for $n \in \mathbf{Z}$.

Proof We will assume that $(*) = \sum_{i=0}^{n} m_o(i)$ is bounded and will obtain a contradiction. By [3,14.11] there is some $f \in C(M)$ with

$$(**) \qquad m(0) = f(\sigma m) - f(m) \quad \text{for all } m \in M .$$

Now let $U = \{x \in X : f$ assumes more than one value on $\pi^{-1}(x)\}$ Since f may be taken to be integer-valued, U is closed. Notice that F is just the set of points x in X such that $\{m(0) : m \in \pi^{-1}(x)\}$ contains more than one point. Thus $x_1 \in U$ since $\pi^{-1}(x_1)$ is one point by assumption 3. By assumption 4, $\{m(0) : m \in \pi^{-1}(z)\}$ is more than one point, and so it follows from (**) that z or Tz is in U. We first consider the case $Tz \in U$. It follows from assumption 4 that if $n > 0$ then $\{m(n) : m \in \pi^{-1}(z)\}$ is a single point, and this and (**) imply that $T^n z \in U$ for $n > 0$. Since $\{T^n z : n > 0\}$ is dense in X, this implies that $X = U$, a contradiction. An analogous argument yields a contradiction in case $z \in U$, so we have shown that no such f exists, and the proof is completed.

Corollary <u>Let</u> $\mathbb{K}$ <u>denote</u> $[0,1)$ <u>considered as the compact group of reals mod</u> 1, <u>and pick an irrational</u> $\alpha \in \mathbb{K}$ <u>and</u> $0 \neq \beta \in \mathbb{K}$. <u>Fix</u> $\gamma, \gamma' \in \mathbb{K}$ <u>and set</u> $A = [\gamma, \gamma+\beta)$, $B = [\gamma', \gamma'+\beta)$. <u>Then</u>

$$N(n) = \sum_{i=0}^{n} \chi_A(i\alpha) - \chi_B(i\alpha)$$

<u>is bounded for</u> $n \in \mathbb{Z}$ <u>iff</u> $\beta \in \mathbb{Z}\alpha$ <u>or</u> $\gamma-\gamma' \in \mathbb{Z}\alpha$.

<u>Proof</u> Notice that for $a,b \in \mathbb{K}$, $[a,b) = \{c \in \mathbb{K} : a \leq c < b\}$ if $a \leq b$ and $[a,b) = [a,1) \cup [0,b)$ if $b < a$. Define $T: \mathbb{K} \to \mathbb{K}$ by $T(x) = x + \alpha$. If $\gamma-\gamma' = k\alpha$ then $A = T^k B$ so $|N(n)| \leq |k|$ for all n. Similarly, if $\beta = k\alpha$ then since $\chi_{[\gamma,\gamma')} - \chi_{[\gamma+\beta,\, \gamma'+\beta)} = \chi_A - \chi_B$ and $T^k[\gamma,\gamma') = [\gamma+\beta, \gamma'+\beta)$ we again have $|N(n)| \leq |k|$. Now we suppose that $\beta \notin \mathbb{Z}\alpha$ and $\gamma-\gamma' \notin \mathbb{Z}\alpha$ and will prove $N(n)$ unbounded. If either

$\gamma-\gamma'-\beta \notin \mathbb{Z}^{\phi}\alpha$ or $\gamma'-\gamma-\beta \notin \mathbb{Z}^{\phi}\alpha$, where $\mathbb{Z}^{\phi} = \mathbb{Z} - \{0\}$, then we can apply the theorem, with $(X, T) = (\mathbf{K}, T)$ and $x_o = 0$, to conclude that $N(n)$ is unbounded. In this case the conditions of the theorem are readily verified except condition 3, and for that choose $z = \gamma$ if $\gamma-\gamma'-\beta \notin \mathbb{Z}^{\phi}\alpha$ and $z = \gamma'$ if $\gamma'-\gamma-\beta \notin \mathbb{Z}^{\phi}\alpha$. The cases $\gamma-\gamma'-\beta \in \mathbb{Z}^{\phi}\alpha$ and $\gamma'-\gamma-\beta \in \mathbb{Z}^{\phi}\alpha$ require a similar but somewhat more detailed analysis, which is done in [2]. The proof is completed.

A slightly weaker form of the corollary was first proved independently by Veech and Furstenberg. The method used in all these proofs, and often in the literature ([6], [5], [4]), namely to use minimality (assumption 1) and a jump discontinuity (assumption 4) to show that a function must be nowhere continuous (thus violating assumption 3), was probably first used by Kakutani [1, page 506] to study the solutions of the eigenfunction equation, which is similar to our formula (**) .

1. H. Furstenberg, The Structure of Distal Flows, Amer. J. Math. 85 (1963), 477 - 515 .

2. H. Furstenberg, H. Keynes, and L. Shapiro, Prime Flows in Topological Dynamics, in preparation.

3. W. Gottschalk and G. Hedlund, Topological Dynamics, Amer. Math. Soc. Colloq. Publ. Vol. 36, Providence, 1955 .

4. K. Petersen and L. Shapiro, Induced Flows, to appear in Trans. Amer. Math. Soc., 1972 .

5. W. Veech, Strict Ergodicity in Zero-dimensional Dynamical Systems and the Kronecker-Weyl Theorem mod 2, Trans. Amer. Soc. 140 (1969), 1 - 33 .

6. H. Furstenberg, Strict Ergodicity and Transformations of the Torus, Amer. J. Math. 83 (1961), 573-601.

University of Minnesota
Minneapolis, Minnesota 55455

MINIMAL SETS AND SOUSLIN SETS

William A. Veech*

1. Introduction.

Certain function spaces which arise in the study of almost periodic functions or topological dynamics have "translation number" definitions, while certain others, as yet, do not. Almost periodic, almost automorphic, and minimal functions fall into the first category, while distal and "point-distal" functions and universal minimal algebras fall into the second. We will see in Section 2 that if Γ is a countable group, then each of the spaces with a translation number definition is a Borel subset of the polonais space, C^Γ, of all complex-valued functions on Γ. On the other hand we will prove in Section 3 that if Γ is infinite and countable, then every universal minimal algebra on Γ fails to be a Souslin subset of C^Γ, let alone a Borel set. Thus it is unlikely that universal minimal algebras can be defined in terms of anything resembling translation numbers. The question of measurability or non-measurability of the distal and point-distal functions (to say nothing of the existence of a translation number definition for either of these spaces) seems to be more difficult, and at present we have only the partial result, proved in Section 5: If Γ is a countable group and α a countable ordinal, the spaces $\mathcal{D}_\alpha$ and $\mathcal{L}_\alpha$ of distal and point-distal functions of order $\leq \alpha$ are Souslin subsets of C^Γ.

---

* Alfred P. Sloan Fellow. Research supported by NSF-GP-18961.

2. Spaces with translation number definitions.

We begin with some notation and definitions. Γ is an infinite group, later to be assumed countable, and $\ell^\infty = \ell^\infty(\Gamma)$ is the Banach space of bounded complex-valued functions on Γ with the sup norm. With the topology of C^Γ ℓ^∞ is σ-compact and hence Borel.

A set $E \subseteq \Gamma$ is said to be left relatively dense if there exist $t > 0$ and $\gamma_1, \ldots, \gamma_t \in \Gamma$ such that $\Gamma = \bigcup_{i=1}^{t} \gamma_i E$. Given $f \in \ell^\infty$, $S \subseteq \Gamma$, and $\varepsilon > 0$ define a set $C_\varepsilon(S) = C_\varepsilon(S,f)$ in Γ by

$$C_\varepsilon(S) = \{\tau \in \Gamma \mid \sup_{s \in S} |f(s\tau) - f(s)| < \varepsilon\} .$$

We say f is right minimal if $C_\varepsilon(S)$ is left relatively dense for all $\varepsilon > 0$ and finite sets S; right almost periodic if $C_\varepsilon(\Gamma)$ is left relatively dense for all $\varepsilon > 0$; and right almost automorphic if it is right minimal, and if for every $\varepsilon > 0$ and S finite there exists $\delta > 0$ and T finite such that $C_\delta(T)^{-1} C_\delta(T) \subseteq C_\varepsilon(S)$.

Each of the spaces just defined has other definitions, often preferred, which are not in terms of translation numbers. We have singled out the translation number definitions for the purpose of the proof of the following proposition. We denote by $\mathfrak{m} = \mathfrak{m}(\Gamma)$, $\mathfrak{a} = \mathfrak{a}(\Gamma)$, and $\mathfrak{a}_a = \mathfrak{a}_a(\Gamma)$ the spaces of right minimal, almost periodic, and almost automorphic functions.

2.1. Proposition.

If Γ is a countable group, the spaces $\mathfrak{m}$, $\mathfrak{a}$, and $\mathfrak{a}_a$ are Borel subsets of C^Γ.

Proof: Let $\mathcal{S}$ be the set of finite subsets of Γ. Fix $s,t,\sigma \in \Gamma$ and $k > 0$, and define a Borel set $\Lambda(s,t,\sigma,k)$ by

$$\Lambda(s,t,\sigma,k) = \{f \,\big|\, |(st^{-1}\sigma) - f(s)| < \tfrac{1}{k}\}$$

It can then be checked that

$$\mathfrak{m} = \ell^\infty \cap \bigcap_{S \in \mathcal{S}} \bigcap_{k=1}^{\infty} \bigcup_{T \in \mathcal{S}} \bigcap_{\sigma \in \Gamma} \bigcup_{t \in T} \bigcap_{s \in S} \Lambda(s,t,\sigma,k) \qquad (2.1)$$

with a similar representation for $\mathfrak{G}$ (delete the intersection over $S \in \mathcal{S}$ and replace S by Γ in (2.1)). As for $\mathfrak{G}_a$, the definition implies

$$\mathfrak{G}_a = \mathfrak{m} \cap \bigcap_{S \in \mathcal{S}} \bigcap_{k=1}^{\infty} \bigcup_{T \in \mathcal{S}} \bigcup_{\ell=1}^{\infty} \bigcap_{t,t' \in \Gamma} (\Lambda_1 \cup \Lambda_2 \cup \Lambda_3)$$

where

$$\Lambda_1 = \bigcup_{\lambda \in T} \{f \,\big|\, |f(\lambda t) - f(\lambda)| \geq \tfrac{1}{\ell}\}$$

Λ_2 is the same with t′ replacing t, and

$$\Lambda_3 = \bigcap_{\sigma \in \mathcal{S}} \{f \,\big|\, |f(\sigma t^{-1} t') - f(\sigma)| < \tfrac{1}{k}\}$$

3. Universal minimal algebras.

We first recall the definition of a "shift operator" or "right shift operator" on $\ell^\infty(\Gamma)$ [5]. We use $M = M(\Gamma)$ to denote the Stone-Cech compactification of the (discrete) group Γ. If $m \in M$ and $f \in \ell^\infty$, $m(f)$ is the value of m at f (or vice-versa). Setting $f_\gamma(t) = f(\gamma t)$, $f \in \ell^\infty$, $\gamma, t \in \Gamma$, we define an operator $T = T_m : \ell^\infty \to \ell^\infty$ by

$$(3.1) \qquad (Tf)(\gamma) = m(f_\gamma)$$

If $\{\gamma_\nu\}$ is a net in Γ convergent to m, then $(Tf)(\gamma) = \lim_\nu f(\gamma\gamma_\nu)$ for all f and γ, for which reason T is called a (right) *shift operator*. Equation (3.1) places M in one-to-one correspondence with the semigroup of all Banach algebra *-endomorphisms of ℓ^∞ which commute with left translation and send the function 1 to itself. Multiplication is left continuous in the M-topology.

A shift operator is *minimal* if $T\ell^\infty \subseteq \mathfrak{m}$ and *minimal idempotent* if it is minimal and $T^2 = T$. Minimal idempotents are well known to exist [2,5], and to each minimal idempotent T_m we associate its range $\mathfrak{m}_m = \{f \in \ell^\infty | T_m f = f\}$. While $\mathfrak{m}$ is generally not a vector subspace of ℓ^∞ (never if Γ is countably infinite), $\mathfrak{m}_m$ is a Γ-subalgebra of $\mathfrak{m}$ (closed, conjugation closed, left invariant, containing the constants).

If A is a minimal Γ algebra (Γ algebra of minimal functions), then every shift operator is isometric on A [5]. In particular, if T_m is a minimal idempotent, $T_m A \subseteq \mathfrak{m}_m$ is isometrically *-isomorphic to A by an isomorphism which commutes with left translation. For this reason $\mathfrak{m}_m$ is said to be a <u>universal minimal algebra</u>. Every universal minimal algebra (= minimal algebra with the stated universal property) is $\mathfrak{m}_m$ for some minimal idempotent T_m, and all are isomorphic [2,5].

The idea for the proof of the following proposition is taken in part from an argument in Ryll-Nardzewski [8]. We assume Γ is countable and <u>infinite</u>.

3.1. <u>Proposition</u>.

For every minimal idempotent, T_m, $\mathfrak{m}_m$ fails to be a Souslin subset of $\mathbf{C}^\Gamma$.

Proof: By a theorem of Ellis [3] $\mathfrak{m}_m$ separates points on Γ, and therefore there is a Γ algebra $B \subseteq \mathfrak{m}_m$ which separates points. We think of B as being identified with $C(X), X = M(B)$ = maximal ideal space of B, and we let (Γ,X) be the minimal flow on X which is adjoint to left translation on B. (See [5] for the minimality of (Γ,X).) Let $x_0 \in X$ correspond to evaluation at $e \in \Gamma$, the identity of Γ. Since B is separable and Γ is infinite, there exists a sequence $\{\gamma_k\}$ in Γ such that $\gamma_k \neq e, \gamma_\ell$, all $k, \ell \neq k$, and $\lim_k \gamma_k x_0 = x_0$.

Now there exists $F \in C(X)$ such that $F(\gamma_k x_0) = \frac{1}{k}$, all k, and $0 < F(X) \leq 1$, $x \in X - \{x_0\}$. (X is metrizable.) Choose a sequence $\{\beta_k\}$ of real numbers such that $\beta_k > \frac{1}{k} > \beta_{k+1}$ for all k, and such that $F(\gamma x_0) \neq \beta_k$ for all γ, k. Define $E_k \subseteq \Gamma$ by $E_k = \{\gamma \mid \beta_{k+1} < F(\gamma x_0) < \beta_k\}$. If φ_k is the characteristic function of E_k for each k, then $\varphi_k \in \mathfrak{m}_m$ and by construction $\varphi_k \varphi_\ell \equiv 0$, $k \neq \ell$, $\sum_{k=1}^{\infty} \varphi_k(\gamma) = 1$, $\gamma \neq e$.

Let N be the natural numbers and 2^N the set of all subsets of N, viewed as the set of 0-1 valued functions $\varepsilon = (\varepsilon_1, \varepsilon_2, \ldots)$. We define a mapping $\pi: 2^N \to \ell^\infty$ by $(\pi\varepsilon)(\gamma) = \sum_{k=1}^{\infty} \varepsilon_k \varphi_k(\gamma)$. If $\gamma = e$, then $\pi\varepsilon(\gamma) = 0$, while if $\gamma \in E_k$, $\pi\epsilon(\gamma) = \epsilon_k$. Let $\{\gamma_\nu\}$ converge to m as before, and fix $\gamma \in E_k$. Since $T_m\varphi_k = \varphi_k$, we have $\gamma\gamma_\nu \in E_k$ for large ν, and therefore by the observation made in the preceding paragraph, $T_m\pi\epsilon = \pi\epsilon$ on $\Gamma - \{e\}$.

Since $\mathfrak{m}_m$ is a vector space of functions containing no function with finite support, there is for each $\epsilon \in 2^N$ at most one value $\delta = 0$ or 1 such that $f \in \mathfrak{m}_m$ where $f(e) = \delta$ and otherwise $f = \pi\epsilon$. On the other hand the last paragraph tells us there is at least one value, namely $\delta = (T_m\pi\epsilon)(e)$. We denote the correspondence by $\delta = \mu(\epsilon)$. $\mu: 2^N \to \mathbf{Z}_2$ is a homomorphism of Boolean rings which is 0 for every finite set (because $\varphi_k \in \mathfrak{m}_m$ and $\varphi_k(e) = 0$), and assigns "mass" 1 to $N = (1,1,\ldots)$. By a theorem of Sierpinski [9] μ cannot be a Borel function on 2^N, because it would then have to be a point mass. We will now prove, using Ryll-Nardzewski's argument, that if $\mathfrak{m}_m$ is a Souslin subset of $\mathbf{C}^\Gamma$, then the graph of μ is a Souslin subset of $2^N \times \mathbf{Z}_2$, meaning that μ must be Borel [1]. From this contradiction our result will follows.

If $\mathfrak{m}_m$ is Souslin, then so is the product space $\mathfrak{m}_m \times 2^N \times \mathbf{Z}_2$. Let Δ be the closed (hence Souslin) subset of this product consisting of triples (f,ϵ,δ) satisfying $f(e) = \delta$ and $f = \pi\epsilon$ on $\Gamma - \{e\}$. For every ϵ there is one and only one choice of f,δ such that $(f,\epsilon,\delta) \in \Delta$, and therefore the graph of μ is the projection of Δ onto the $2^N \times \mathbf{Z}^2$ coordinates. Thus, the graph of μ is Souslin, and as remarked above, the proposition follows.

3.2. <u>Question</u>.

If $m \in M(\Gamma) - \Gamma$, then Sierpinski's theorem implies the set of $E \in 2^\Gamma$ corresponding to neighborhoods of m is not a Souslin subset of 2^Γ. Let J be the set of $E \in 2^\Gamma$ for which there exists <u>some</u> minimal idempotent (or just idempotent) T_m such that E corresponds to a neighborhood of m. Can anything be said about J?

4. <u>Souslin</u> <u>algebras</u>.

Let Γ be a countable <u>set</u>, and suppose A is a closed subalgebra of

$\ell^{\infty}(\Gamma)$ containing the constants. There is a canonical map of Γ into $X = X(A)$, the maximal ideal space of A, and therefore Γ inherits a natural topology, $\mathcal{T}_A$, and a natural uniformity, $\mathcal{U}_A$. Let $C = C[0,1]$, and let Λ be a subset of C. In the present section we consider the spaces $B = B(A,\Lambda)$ and $B_u = B_u(A,\Lambda)$ of $\mathcal{T}_A$ continuous and $\mathcal{U}_A$ uniformly continuous functions from Γ to Λ. We will see that if A is self-adjoint and a Souslin subset of C^{Γ}, and if Λ is a Borel set with compact closure, then both B and B_u are Souslin subsets of the polonais space C^{Γ}. In all that follows we suppose A to be Souslin.

Let R be the rational numbers in $[0,1]$ and set up the product space

$$A^* = \prod_{r \in R} A_r \qquad (A_r = A)$$

which with the product topology is also Souslin. We may and shall regard A^* as a subset of $(C^{\Gamma})^R$. Given $\gamma \in \Gamma$, $k > 0$, and a function g on R, the set

$$(4.1) \qquad \Sigma(\gamma,k,g) = \{(f_r)_{r \in R} \in A^* \mid \sup_{r \in R} |f_r(\gamma) - g(r)| < \tfrac{1}{k}\}$$

is the intersection of a Borel set in $(C^{\Gamma})^R$ with A^* and hence is Souslin. Given a countable set $\{g_n\}$ in $\ell^{\infty}(R)$, define $D = D(\{g_n\})$ by

$$D = \bigcap_{k=1}^{\infty} \bigcap_{\gamma \in \Gamma} \bigcup_{n=1}^{\infty} \Sigma(\gamma,k,g_n)$$

Let $\Lambda \subseteq \ell^{\infty}(R)$ be the norm closure of $\{g_n\}$. If $(f_r)_{r \in R} \in D$, then for each $\gamma \in \Gamma$ and $\epsilon > 0$ there exists an n such that if $F_{\gamma}(r) = f_r(\gamma)$, $\sup_r |F_{\gamma}(r) - g_n(r)| < \epsilon$. Since ϵ, γ are arbitrary, $F_{\gamma}(\cdot) \in \Lambda$ for all γ. Conversely, if $F_{\gamma}(\cdot) \in \Lambda$ for all γ, and if $(f_r)_{r \in R} \in A^*$, where $f_r(\gamma) = F_{\gamma}(r)$, then $(f_r)_{r \in R} \in D$.

In addition to the above let us assume Λ is compact and each g_n is uniformly continuous on R. We may in particular regard Λ as a subset of C. If $(f_r)_{r \in R} \in D$, and if $\{\gamma_\nu\}$ is a net which is $\mathcal{U}_A$ Cauchy, then for every $r \in R$ $\lim_\nu f_r(\gamma_\nu) = \lim_\nu F_{\gamma_\nu}(r)$ exists. Because Λ is compact, the Arzela-Ascoli theorem tells us $F_{\gamma_r}(\cdot)$ is norm convergent. Thus, the function $\gamma \to F_\gamma$ is $\mathfrak{U}_A$ uniformly continuous from Γ to Λ. Conversely, if $\gamma \to F_\gamma$ is $\mathfrak{U}_A$ uniformly continuous, then for each $r \in R$, $f_r(\gamma) = F_\gamma(r)$ is $\mathfrak{U}_A$ uniformly continuous. If A is self-adjoint, which we now assume, then $f_r \in A$, and so $(f_r)_{r \in R} \in D$. We have proved

4.1. Proposition.

If A is a closed, self-adjoint subalgebra of $\ell^\infty(\Gamma)$ containing the constants, if $\Lambda \subseteq C[0,1]$ is a compact set, and if A is a Souslin subset of C^Γ, then $B_u(A,\Lambda)$ is a Souslin subset of C^Γ.

4.2. Corollary.

If $\Lambda' \subseteq \Lambda$ is a Borel set, and if notations and assumptions are as above, then $B_u(A,\Lambda')$ is a Souslin set.

4.3. Remark.

The map from D to B_u is one-to-one, and therefore if A is a Borel set, B_u is also Borel.

In order to study $B(A,\Lambda)$, it is necessary to obtain information about $\mathfrak{J}_A$ as a subset of 2^Γ. No generality will be lost if we assume A is real. The set $A_0 = \{f \in A \mid f(\gamma) \notin \mathbb{Q},\ \gamma \in \Gamma\}$, $\mathbb{Q}$ = rational numbers, is dense in A and therefore determines the same topology on X as does A. Therefore, A_0 determines the same topology on Γ as does A, namely $\mathfrak{J}_A$.

For every integer $n > 0$ and collection (a_i,b_i), $a_i < b_i$, $i = 1,\ldots,n$, of pairs of rational numbers, we define a C^Γ continuous mapping from A_0^n to a subset of $\mathfrak{J}_A$ by

$$T(f_1,\cdots,f_n) = \bigcap_{i=1}^{n} \{\gamma \mid a_i < f(\gamma) < b_i\}$$

A_0 is a Souslin set (intersection of a Souslin set and a Borel set), and therefore the (countable) union $\mathfrak{J}'$ of the ranges of all such maps T is a Souslin subset of $\mathfrak{J}_A \subseteq 2^\Gamma$. Clearly, $\mathfrak{J}'$ is a basis for $\mathfrak{J}_{A_0} = \mathfrak{J}_A$.

For each $\gamma \in \Gamma$ define $S_\gamma = \{E \in 2^N | \gamma \in E\}$ and $\mathfrak{J}'_\gamma = S_\gamma \cap \mathfrak{J}'$. The map $\pi_\gamma: \mathfrak{J}'_\gamma \times 2^N \to 2^N$ defined by $\pi_\gamma(E,F) = E \cup F$ is continuous, and therefore its range, $\mathfrak{h}_\gamma$, is also Souslin. $\mathfrak{h}_\gamma$ is the collection of all $\mathfrak{J}_A$ neighborhoods of γ. This being so, we clearly have

$$\mathfrak{J}_A = \bigcap_\gamma (\mathfrak{h}_\gamma \cup S^c_\gamma)$$

and $\mathfrak{J}_A$ is a Souslin set in 2^Γ. Next, let $\sigma: 2^N \to 2^N$ be $\sigma E = E^c$. σ is continuous, hence $\sigma\mathfrak{J}_A$ is Souslin as is $\mathfrak{J}_A \cap \sigma\mathfrak{J}_A = C_A$, the set of clopen sets in $\mathfrak{J}_A$.

Let $g_1, g_2, \ldots$ be any dense subset of the compact set $\Lambda \subseteq C$, and for each n define $P_n: C^n_A \to B(A,\Lambda)$ by

$$P_n(E_1, \cdots, E_n) = \sum_{i=1}^n g_i \chi_{Fi}$$

where $F_1 = E_1$, $F_{i+1} = E_{i+1} \cap (\bigcup_{j=1}^i E_j)$, $1 \le i < n$. P_n is continuous when $B(A,\Lambda)$ is given the Λ^Γ topology, and so $B' = \bigcup_{n=1}^\infty P_n C^n_A$ is a Souslin subset of $B(A,\Lambda)$ in Λ^Γ. Given $F \in B$ and $\epsilon > 0$ there exists an n such that $g_1, \ldots, g_n$ is an $\epsilon/2$ net in Λ, and then there exists δ, $\epsilon/2 < \delta \le \epsilon$, such that for each $\gamma \in \Gamma$ and i, $1 \le i \le n$, $\|F(\gamma) - g_i\| \neq \delta$. Define $E_i = \{\gamma | \|F(\gamma) - g_i\| < \delta\}$. Then $(E_1, \cdots, E_n) \in C^n_A$ and if $G = P_n(E_1, \cdots, E_n)$, $\|G(\gamma) - F(\gamma)\| < \epsilon$, all γ. Thus, if we define $B''_k = \{H | \|H(\gamma)\| \le 1/k, \gamma \in \Gamma\}$, we see that

$$B(A,\Lambda) = \bigcap_{k=1}^\infty (B' + B''_k)$$

and $B(A,\Lambda)$ is a Souslin set. ($B' + B''_k$ is a continuous image of $B' \times B''_k$, hence Souslin.)

4.4. Proposition.

With notations and assumptions as above $B(A,\Lambda)$ is a Souslin subset of $\Lambda^\Gamma \subseteq C^\Gamma$. In fact, this conclusion is true for any relatively compact Borel set $\Lambda \subseteq C$.

4.5. Theorem. If A is a closed, self-adjoint subalgebra of $\ell^\infty(\Gamma)$ which is also a Souslin subset of C^Γ, and if Λ is a Borel subset of a compact metric space Λ', then $B(A,\Lambda)$ and $B_u(A,\Lambda)$ are Souslin subsets of Λ^Γ.

Proof: Use previous results together with the fact every separable metric space isometrically embeds in C.

5. Relative almost automorphy and relative almost periodicity.

In this section Γ is a countable group, and A is a right minimal Γ algebra. We associate $\mathcal{J}_A$ and $\mathcal{U}_A$ to A as in Section 4. If $G \subseteq C^\Gamma$ and $\gamma \in \Gamma$, G_γ denotes the set of right translates by γ of elements of G: $G_\gamma = \{g^\gamma | g \in G\}$, where $g^\gamma(t) = g(t\gamma)$.

If $0 < M < \infty$, let $S_M \subseteq \ell^\infty$ be the set $\{f \mid \|f\|_\infty \leq M\}$. S_M is compact in C^Γ, and we denote by $\Lambda' = \Lambda'_M$ the set of compact subsets of S_M with the Hausdorff metric topology. It is easy to see that $\Lambda \subseteq \Lambda'$, the set of $E \in \Lambda'$ which are norm compact is a Borel subset of Λ'. Thus, if A is Souslin, the spaces $B(A,\Lambda')$ and $B_u(A,\Lambda')$ are Souslin, by Theorem 4.5.

In preparation of Definition 5.1 define E(A) to be the set of shift operators which act as the identity on A. E(A) is a closed subsemigroup of the set of all shift operators, and because A is minimal, E(A) contains a minimal idempotent [5].

5.1. Definition.

We say $f \in \ell^\infty$ is A-almost automorphic if E(A)f is norm compact, and if $F \in B(A,\Lambda')$, where $F(\gamma) = (E(A)f)_\gamma$.

$\mathcal{H}(A)$ denotes the set of A-almost automorphic functions. Of course $A \subseteq \mathcal{H}(A)$, and it will develop that $\mathcal{H}(A)$ is a minimal Γ-algebra.

Let $H(A,\Lambda')$ be the subset of $B(A,\Lambda')$ consisting of those F such that $F(\gamma) = F(e)_\gamma$, $\gamma \in \Gamma$. The latter is a Borel condition, and therefore H is also Souslin, as is the corresponding space $H_u(A,\Lambda') \subseteq B_u(A,\Lambda')$.

5.2. Lemma.

If $F \in H$, then $T_m F(e) = F(e)$ for all $T_m \in E(A)$.

Proof: Let $\{\gamma_\nu\}$ be a net in Γ which is convergent to m. If $g \in F(e)$, the $\mathcal{T}_A$ continuity of F implies there exists a net $\{g_\nu\}$ in F(e) such that $\{(g_\nu)^{\gamma_\nu}\}$ is convergent to g in C^Γ. Since F(e) is norm compact, we may suppose $\{g_\nu\}$ is norm convergent to some $h \in F(e)$. Then obviously $\{h^{\gamma_\nu}\}$ converges to g, meaning $g = T_m h \in T_m F(e)$. Since g is arbitrary, $T_m F(e) = F(e)$, as claimed.

5.3. Lemma.

If $F \in H$, then $F(e) \subseteq \mathfrak{m}_m$ for some minimal idempotent T_m.

Proof: Let $T_m \in E(A)$ be a minimal idempotent. If $g \in F(e)$, then $g = T_m h$, some $h \in F(e)$, and therefore $T_m g = g$. That is, $g \in \mathfrak{m}_m$.

5.4. Lemma. If $f \in \ell^\infty$ is such that $E(A)f$ is norm compact and contained in $\mathfrak{m}_m$ for some minimal idempotent T_m, then f is A-almost automorphic.

Proof: Since E(A) is a semigroup, we have $T_n E(A)f \subseteq E(A)f$, $T_n \in E(A)$. Since $E(A)f \subseteq \mathfrak{m}_m$, T_n is isometric on E(A)f, and since an isometry of a compact metric space is always surjective, $T_n E(A)f = E(A)f$, $T_n \in E(A)$. From this it follows readily that $\gamma \to (E(A)f)_\gamma$ is $\mathcal{T}_A$ continuous at e. The extension to arbitrary γ is straightforward, and therefore $f \in \mathcal{H}(A)$.

5.5. Proposition.

A necessary and sufficient condition that $f \in \ell^\infty$ be A-almost automorphic and bounded by M is that there exist $F \in H$ with $f \in F(e)$.

Proof: Necessity is proved by setting $F(\gamma) = (E(A)f)_\gamma$. For sufficiency we note that $E(A)f \subseteq F(e)$ is norm compact, and because $F(e) \subseteq \mathfrak{m}_m$, some m, $E(A)f \subseteq \mathfrak{m}_m$. The proposition follows by Lemma 5.4.

5.6. Lemma.

Let L be a Souslin subset of the set of all closed subsets of a compact metric space X, and let L_0 be the set of $x \in X$ for which there exists $\ell \in L$ with $x \in \ell$. Then L_0 is a Souslin subset of X.

Proof: By definition there exists a polonais space Y and a continuous surjection $\varphi: Y \to L$. Define $\Delta \subseteq Y \times X$ to be the set $\Delta = \{(y,x) \mid x \in \varphi(y)\}$. Δ is closed, and L_0 is just the projection of Δ onto the x coordinate.

5.7. Theorem.

If A is a minimal Γ-algebra, then $\aleph(A)$ is also a minimal Γ-algebra. If A is Souslin in C^Γ, then $\aleph(A)$ is also Souslin.

Proof: If $f,g \in \aleph(A)$, then $E(A)fg \subseteq E(A)fE(A)g$, $E(A)(f+g) \subseteq E(A)f + E(A)g$, and so $E(A)fg$, $E(A)(f+g)$, which are closed subsets of C^Γ, are compact. We know for any $T_m \in E(A)$ which is minimal idempotent that T_m is the identity on $\aleph(A)$. Thus, $\aleph(A) \subseteq \mathfrak{m}_m$, and so $fg, g+g \in \aleph(A)$. As for the second assertion, Lemma 5.6 tells us $S_M \cap \aleph(A) = \bigcup_{F \in H} F(e)$ is Souslin, and therefore letting $M \to \infty$ through a sequence, $\aleph(A)$ is also Souslin.

5.8. Definition.

We say $f \in \ell^\infty$ is A-almost periodic if $T_n f$ is $T_n A$-almost automorphic for every $n \in M(\Gamma)$.

5.9. Remark.

It is not difficult to see that the definition just given is equivalent, in the context of distal functions, to the definition given in [6]. Also, it is easy to check that our definition is equivalent to the definition arising in 5.1 from replacing "$\mathfrak{U}_A$ uniformly continuous" for "$\mathfrak{J}_A$ continuous". We omit the details, but this leads to

5.10. Theorem.

If A is a minimal Γ-algebra, then $\aleph_u(A)$, the space of A-almost periodic functions, is a minimal Γ-algebra. If A is Souslin, then $\aleph_u(A)$ is also Souslin.

5.11. Definition.

We define for each ordinal $\alpha < \Omega$ an algebra $\mathcal{D}_\alpha = \mathcal{D}_\alpha(\Gamma)$ by

$$\mathcal{D}_0 = \text{constants}$$

$$\mathcal{D}_{\alpha+1} = \mathcal{N}_u(\mathcal{D}_\alpha)$$

$$\mathcal{D}_{\lim \beta<\alpha} = \text{closure} \bigcup_{\beta<\alpha} \mathcal{D}_\beta \qquad (\alpha = \text{limit ordinal})$$

$\mathcal{D}_\alpha$ is the space of distal functions of order α. Next, define $\mathcal{L}_\alpha$, $\alpha < \Omega$, in the same way, using $\mathcal{N}(\mathcal{L}_\alpha)$ in place of $\mathcal{N}_u(\mathcal{L}_\alpha)$.

5.12. Theorem.

For each $\alpha < \Omega$ both $\mathcal{D}_\alpha$ and $\mathcal{L}_\alpha$ are minimal Γ algebras which are Souslin subsets of C^Γ.

Proof: Use Theorems 5.7 and 5.10, the fact the norm closure of a Souslin set is Souslin, and transfinite induction.

5.13. Remark.

Recall that $\mathcal{L} = \mathcal{L}(\Gamma)$ is the set of f such that $T_m f = f$, all minimal idempotents T_m [7]. If the structure theorem of [11] is true without the "residuality of distal points assumption," then $\mathcal{L}_0 = \bigcup_\alpha \mathcal{L}_\alpha$ will be universal for $\mathcal{L}$ in the sense that for any separable set $B \subseteq \mathcal{L}$ there is a shift operator T_n with $T_n B \subseteq \mathcal{L}_0$. The nature of the inclusion $\mathcal{L}_0 \subseteq \mathcal{L}$ is not understood.

$\mathcal{D} = \mathcal{D}(\Gamma)$ is the set of $f \in \ell^\infty$ such that $T_n f \in \mathcal{L}$ for all $n \in M(\Gamma)$. It is known for countable Γ that $\mathcal{D} = \bigcup_\alpha \mathcal{D}_\alpha$ [4,6].

5.14. Remark.

Define $\pi : \ell^\infty \times \mathfrak{m} \to \ell^\infty$ by $\pi(f,g) = f + g$. $\mathcal{L}$ is characterized as the set of f such that $\pi(f,g) \in \mathfrak{m}$ for all $g \in \mathfrak{m}$. Thus $\mathcal{L}^c \subseteq \ell^\infty$ is the projection onto the f coordinate of $\pi^{-1}\mathfrak{m}$ (a Borel set because $\mathfrak{m}$ is Borel) and therefore $\mathcal{L}^c$ is Souslin.

Rice University

BIBLIOGRAPHY

[1] Bourbaki, N. Topologie Generale, Chap. 9, Paris, Hermann, 1958.

[2] Ellis, R. "A semigroup associated with a transformation group", T.A.M.S., v. 94 (1960), 272-281.

[3] ________. "Universal minimal sets", P.A.M.S., v. 11 (1960), 540-543.

[4] Furstenberg, H. "The structure of distal flows", Am. J. of Math., v. 85 (1963), 477-515.

[5] Knapp, A. W. "Decomposition theorem for bounded uniformly continuous functions on a group", Am. J. of Math., v. 88 (1966), 901-914.

[6] ________. "Distal functions on groups", T.A.M.S., v. 128 (1967), 1-40.

[7] ________. "Functions behaving like almost automorphic functions", Topological Dynamics, An International Symposium, Ed. by J. Auslander and W. Gottschalk, N. Y., W. A. Benjamin, 299-317.

[8] Ryll-Nardzewski, C. "Concerning almost periodic extensions of functions", Colloq. Math., 12(1964), 235-237.

[9] Sierpinski, W. "Fonctions additives non complètement additives et fonctions non mesurables", Fund. Math., 33 (1938), 96-99.

[10] Veech, W. A. "Almost automorphic functions on groups", Am. J. of Math., v. 87 (1965), 719-751.

[11] ________. "Point distal flows", Am. J. of Math., v. 92 (1970), 205-242.

[12] ________. "G-almost automorphic functions", Problems in Analysis, 345-351.

SOME RESULTS ON THE CLASSIFICATION OF NON-INVERTIBLE MEASURE PRESERVING TRANSFORMATIONS

Peter Walters

In this talk I shall describe some of the ideas in the paper [4] (joint work with W. Parry) and illustrate them with examples.

Throughout all measure spaces will be Lebesgue spaces, (see [5]) $(X, \mathcal{B}, m)$ will be the usual notation and $T : X \to X$ will denote a measure-preserving transformation (not necessarily invertible). If $(X_1, \mathcal{B}_1, m_1)$, $(X_2, \mathcal{B}_2, m_2)$ are Lebesgue spaces and $T_1 : X_1 \to X_1$, $T_2 : X_2 \to X_2$ are measure-preserving then they are isomorphic if there exists (i) $\underline{X}_1 \in \mathcal{B}_1$ with $m_1(\underline{X}_1) = 1$ and $T_1 \underline{X}_1 \subset \underline{X}_1$,
(ii) $\underline{X}_2 \in \mathcal{B}_2$ with $m_2(\underline{X}_2) = 1$ and $T_2 \underline{X}_2 \subset \underline{X}_2$,
(iii) an invertible measure-preserving transformation $\varphi : \underline{X}_1 \to \underline{X}_2$ such that $\varphi T_1(x) = T_2 \varphi(x) \quad x \in \underline{X}_1$. When this holds we shall write $T_1 \cong T_2$ and when we want to emphasise the isomorphism φ we shall write $T_1 \overset{\varphi}{\cong} T_2$. We shall sometimes call a measure-preserving transformation $T : X \to X$ an endomorphism : homomorphism, automorphism and isomorphism will be used in the usual way. The classification problem in Ergodic Theory is to determine when two given endomorphisms T_1, T_2 are isomorphic. The usual way to approach such problems is to find invariants of the equivalence relation $(\cong)$ and then see how good these invariants are. We shall consider some invariants of the equivalence relation $\cong$ that are trivial for automorphisms.

ε will denote the partition of X into points, ν will denote the trivial partition of X and $\mathcal{N}$ will denote the trivial sub σ-algebra of $\mathcal{B}$. We shall always consider countable-to-one endomorphisms. This assumption is not needed for most of the discussion but is sometimes convenient. Since $h(T) \geq H(\varepsilon / T^{-1}\varepsilon)$ every endomorphism with finite entropy is countable-to-one. If $T : X \to X$ is countable-to-one then T is isomorphic to a positively measurable and positively non-singular transformation [7]. So we may as well assume that all our endomorphisms are positively measurable ($B \in \mathcal{B} \Rightarrow TB \in \mathcal{B}$) and positively non-singular ($m(B) = 0 \Rightarrow m(TB) = 0$). Essentially this means that each element of the partition $T^{-1}\varepsilon$ consists of a countable number of points each having positive canonical measure ($m(x \mid T^{-1}Tx) > 0$).

*

(1) The index.

If $T : X \to X$ is an endomorphism the index $i_T : X \to Z^+$ is defined by $i_T(x)$ = the number of atoms of the measure $m(\ \mid T^{-1}x)$. With our assumption on T $i_T(x)$ = the cardinality of $T^{-1}x$. If $T_1 \overset{\varphi}{\cong} T_2$ then $i_{T_1}(x) = i_{T_2}(\varphi x)$ a.e. Hence isomorphic transformations have isomorphic index functions.

*

(2) The jacobian.

This was first introduced in Parry's book [3]. The idea is to capture how T expands measure locally. Consider the partition $T^{-1}\varepsilon$ and let $m(\ \mid T^{-1}Tx)$ denote the canonical measure in the fibre containing x. Define the jacobian of T as $j_T(x) = \dfrac{1}{m(x \mid T^{-1}Tx)}$.
(If T is not countable-to-one this definition still makes sense if we put $j_T(x) = \infty$ when $m(x \mid T^{-1}Tx) = 0$). Another way of defining j_T is as follows. By a theorem of Rohlin [5] we can partition X into measurable sets $\{A_1, A_2, \ldots\}$ which are cross-sections of $T^{-1}\varepsilon$. Then $T|_{A_i}$ is one-to-one and $B \to m(T|_{A_i}(B \cap A_i))$ defines a measure on X which is absolutely continuous with respect to m. If $\frac{dm(T|_{A_i})}{dm}$ denotes the Radon-Nikodym derivative then $j_T(x) = \sum_i \chi_{A_i}(x) \frac{dm(T|_{A_i})}{dm}(x)$. It is clear from this expression that j_T measures how T expands the measure locally.

(a) T is an automorphism $\Leftrightarrow$ $j_T = 1$ a.e.

(b) It makes sense to define j_ψ for a homomorphism $\psi : X_1 \to X_2$. If $T_1 \overset{\varphi}{\cong} T_2$ then $j_{T_1}(x) = j_{T_2}(\varphi x)$ a.e. Hence isomorphic transformations have isomorphic jacobians. We shall write $j_{T_1} \cong j_{T_2}$ when $j_{T_1}(x) = j_{T_2}(\varphi x)$ a.e. for some isomorphism $\varphi : X_1 \to X_2$.

(c) $\int_X \log j_T(x)\, dm(x) = H(\varepsilon / T^{-1}\varepsilon)$. So if $H(\frac{\varepsilon}{T^{-1}\varepsilon}) = h(T)$, which

will be true if T has a 1-sided generator with finite entropy, then
$h(T) = \int_X \log j_T(x)\, dm(x)$.

Examples.

(i) Bernoulli shifts.

Suppose the states are 0, 1, 2, 3,.....with weights $p_o, p_1, p_2, \ldots,$. (If the state space is finite with k elements then put $0 = p_k = p_{k+1} = \ldots$) If $A_i = \{ \{x_n\}_{n=o}^{\infty} \mid x_o = i\}$ then $A_o, A_1, A_2, \ldots,$ partition X and are cross-sections of $T^{-1}\varepsilon$. Hence
$j_T(x) = \sum_{i=0}^{\infty} \chi_{A_i}(x) \frac{1}{p_i}$.

THEOREM 1.

Two 1-sided Bernoulli shifts T_1, T_2 are isomorphic <=> their jacobians are isomorphic <=> their state spaces are isomorphic.

Proof.

Suppose T_1 has states 0, 1, 2,... with weights $p_o \geq p_1 \geq p_2 \geq \cdots$ and let T_2 have states 0, 1, 2,... with weights $q_o \geq q_1 \geq q_2 \geq \ldots$. Suppose $j_{T_1} \cong j_{T_2}$. This means that j_{T_1} and j_{T_2} assume the same value on sets of equal measure. One then easily concludes that $p_o = q_o$ and then that $p_1 = q_1$, and so on.

For the case of 2-sided Bernoulli shifts D. Ornstein has shown that entropy is a complete invariant. The above shows one needs the function j_T rather than the number h(T) for the 1-sided case. Recall from (c) that $h(T) = \int_X \log j_T(x)\, dm(x)$.

Theorem 1 can easily be extended to the case of a general (not necessarily countable) state space.

(ii) 2-state Markov shifts.

We shall consider the mixing case. The general transition matrix is $\begin{bmatrix} p & 1-p \\ p' & 1-p' \end{bmatrix}$ $\quad 0 \leq p \leq 1, \quad 0 \leq p' \leq 1$
and the invariant initial distribution is $\left(\frac{p'}{1-p+p'}, \frac{1-p}{1-p+p'}\right)$.

Let $A_{ij} = \left\{ \{x_n\}_{n=0}^{\infty} \mid x_0 = i,\ x_1 = j \right\}$ $i, j = 0, 1$. If we use the binary expansion of a number in $I = [0, 1)$ we can transfer the measure to I and T becomes $Tx = 2x \bmod 1$. The sets A_{ij} are then as in the diagram

A_{00}	A_{01}	A_{10}	A_{11}
0 – $1/4$	$1/4$ – $\frac{1}{2}$	$\frac{1}{2}$ – $3/4$	$3/4$ – 1

and the value of j_T is constant on the sets A_{ij} and is given by

value of j_T	$\frac{1}{p}$	$\frac{1}{p'}$	$\frac{1}{1-p}$	$\frac{1}{1-p'}$
measure	$\frac{pp'}{1-p+p'}$	$\frac{p'(1-p)}{1-p+p'}$	$\frac{p'(1-p)}{1-p+p'}$	$\frac{(1-p)(1-p')}{1-p+p'}$

In this case of 2-state mixing Markov shifts the jacobian does not determine isomorphism. There are non-isomorphic examples with isomorphic jacobians. (see example (vi)).

(iii) Toral endomorphisms.

Let $A : K^n \to K^n$ be a group-endomorphism of the n-torus given by the nxn matrix $[A]$ of integers with $\det [A] \neq 0$. $A^{-1}\varepsilon$ is the partition into cosets of kernel(A) and so each set in $A^{-1}\varepsilon$ has $|\det[A]|$ members. Hence, since all points have the same conditional measure, $j_T \equiv |\det[A]|$. The jacobian does not determine isomorphism in this case. Let $T_1\binom{x}{y} = \binom{2x}{2y} \bmod 1$. $j_{T_1} \equiv 4$. Let φ be the automorphism $\varphi\binom{x}{y} = \begin{pmatrix} 2 & 1 \\ 1 & 1 \end{pmatrix} \binom{x}{y}$ and put $T_2 = T_1\varphi = \varphi T_1$. $j_{T_2} \equiv 4$. However $h(T_1) \neq h(T_2)$ so $T_1 \not\cong T_2$. Both T_1 and T_2 are exact ($\bigwedge_0^{\infty} T^{-n}\varepsilon = \nu$). In fact T_2 is not isomorphic to a Markov shift because it does not have a 1-sided generator with finite entropy (this is because $h(T_2) > H(\frac{\varepsilon}{T_2^{-1}\varepsilon})$). Suppose that A is an ergodic (group) endomorphism of K^n with a generator $(h(A) = H(\frac{\varepsilon}{A^{-1}\varepsilon}))$ then all the eigenvalues of $[A]$ have absolute value greater than 1 and also A is exact. It seems reasonable to conjecture in this case that A is isomorphic to the 1-sided Bernoulli shift on $|\det(A)|$ points with equal weights.

———— × ————

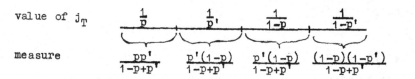

(3) The σ-algebra $\beta(T)$.

For an endomorphism $T : X \to X$ let $\beta(T)$ denote the smallest σ-algebra such that $T^{-1}\beta(T) \subseteq \beta(T)$ and j_T is measurable with respect to $\beta(T)$. i.e. $\beta(T) = \bigvee_{n=0}^{\infty} T^{-n} j_T^{-1}(\mathcal{L})$. $\mathcal{L}$ = Lebesgue sets in R. If $T_1 \overset{\varphi}{\cong} T_2$ then $\beta(T_2) = \varphi\beta(T_1)$ and the quotient transformations $T_{\beta(T_1)}$, $T_{\beta(T_2)}$ are isomorphic. Therefore isomorphic endomorphisms have isomorphic $\beta(T)$'s.

Examples.

(iv) $(\frac{1}{2}, \frac{1}{2})$-shift.

We know by example (i) that $j_T \equiv 2$ so $\beta(T) = \mathcal{N}$.

(v) (p, q)-shift $p \neq \frac{1}{2}$.

By example (i) $j_T(x) = \begin{cases} 1/p & \text{if } x_o = o \\ 1/q & \text{if } x_o = 1 \end{cases}$, so $\beta(T) = \mathcal{B}$.

(vi) We shall use $\beta(T)$ to prove the assertion made in example (ii). Let T_1 be the (p, q) shift, $p \neq q$. We can consider T_1 as the Markov shift with transition matrix $\begin{bmatrix} p & q \\ p & q \end{bmatrix}$. Let T_2 be the Markov shift with transition matrix $\begin{bmatrix} p & q \\ q & p \end{bmatrix}$. Then by example (ii) we can illustrate their jacobians as follows.

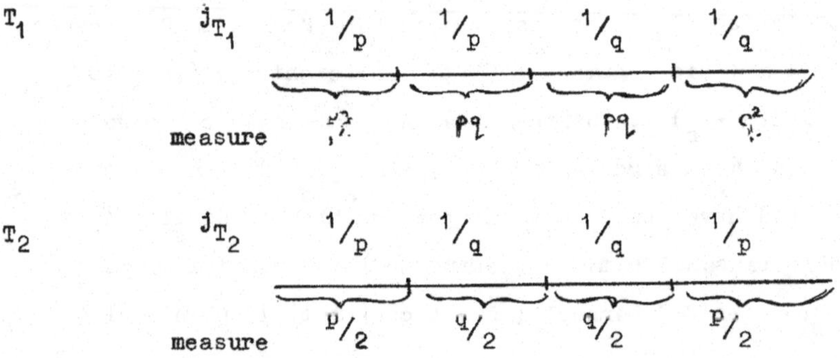

Clearly j_{T_1} is isomorphic to j_{T_2}. We know $\beta(T_1) = \mathcal{B}$ by example (v).

However $\beta(T_2) \neq \mathcal{B}$ because j_{T_2} is invariant under the map $\psi(\{x_n\}) = \{x_n+1 \bmod 1\}$ which commutes with T_2 and therefore x and $\psi(x)$ belong to the same elements of $\beta(T_2)$. H. Furstenberg was the first to mention the non-isomorphism of T_1 and T_2 [1]. We shall show later that there are non-isomorphic 2-state mixing Markov shifts with isomorphic jacobians and isomorphic algebras $\beta(T)$.

THEOREM 2.

If $\beta(T) = \mathcal{B}$ then the only automorphism commuting with T is the identity.

Proof.

If φ is an automorphism with $\varphi T = T\varphi$ then x and $\varphi(x)$ belong to the same elements of $\beta(T) = \mathcal{B}$ and so $x = \varphi(x)$.

COROLLARY

The identity is the only automorphism commuting with the 1-sided (p, q)-shift $(p \neq q)$.

In [4] we compute all automorphisms commuting with the $(\frac{1}{2}, \frac{1}{2})$ shift. This is done during a study of the decreasing sequence of σ-algebras $\{T^{-n}\mathcal{B}\}_0^\infty$. If $T_1 \overset{\varphi}{\cong} T_2$ then $\varphi T_1^{-n}\mathcal{B}_1 = T_2^{-n}\mathcal{B}_2$, $n \geq 0$. Some results on the isomorphism of such sequences are given in [4].

———————— × ————————

One could hope that for exact endomorphisms (those which are the "opposite" of automorphisms – $\bigcap_0^\infty T^{-n}\mathcal{B} = \mathcal{N}$) the invariants i_T, j_T, $\beta(T)$ and $h(T)$, taken together, may be complete. This is not so. We shall give an example to illustrate this and then talk about the new invariant one needs to get around this example. In fact the example is of two Markov shifts on a state space with two points. Before giving the example I will state a theorem whose proof includes the example.

THEOREM 3.

If T_1, T_2 are mixing 1-sided Markov shifts on a state space with two points then $j_{T_1} \cong j_{T_2}$ implies $T_1 \cong T_2$ except in the following cases:-

(1) T_1 has transition matrix $\begin{bmatrix} p & q \\ p & q \end{bmatrix}$ and T_2 has matrix $\begin{bmatrix} q & p \\ p & q \end{bmatrix}$

$p \neq q$. $\beta(T_1)$ is not isomorphic to $\beta(T_2)$ so $T_1 \ncong T_2$.

(2) T_1 has transition matrix $\begin{bmatrix} p & q \\ p & q \end{bmatrix}$ and T_2 has matrix $\begin{bmatrix} p & q \\ q & p \end{bmatrix}$

$p \neq q$. $\beta(T_1)$ is not isomorphic to $\beta(T_2)$ so $T_1 \ncong T_2$.

(3) T_1 has transition matrix $\begin{bmatrix} p & q \\ q & p \end{bmatrix}$ and T_1 has matrix $\begin{bmatrix} q & p \\ p & q \end{bmatrix}$

$p \neq q$. Then $\beta(T_1) = \beta(T_2)$ but $T_1 \ncong T_2$.

Case (2) has been discussed above. In the case of 2-sided mixing Markov shifts Friedman and Ornstein have shown that entropy is a complete isomorphism invariant.

The following example proves part (3) of the theorem.

Let $S:(Y,\mathcal{C},\mu) \to Y$ be the 1-sided (p,q) shift $(p \neq q)$ with state space $\{-1,1\}$. Put $X = Y \times Z_2$ where Haar measure is used on the two element group $Z_2 = \{-1,1\}$. Define $T_1 : X \to X$ by $T_1(\{y_n\}_0^\infty, g) = (S\{y_n\}, y_0 g)$. It is easy to check that T_1 is weak-mixing and then show that T_1 is exact. T_1 is isomorphic to the Markov shift with transition matrix $\begin{bmatrix} p & q \\ q & p \end{bmatrix}$. If $\pi : X \to Y$ is the natural projection then $j_{T_1}(x) = j_S(\pi x)$ and $\beta(T_1) = \pi^{-1}(\mathcal{C})$. If $\phi(\{y_n\},g) = (\{y_n\},-g)$ then $\phi T_1 = T_1 \phi$ and $T_2 = \phi T_1$ has the form $T_2(\{y_n\}_0^\infty, g) = (S\{y_n\}, -y_0 \cdot g)$. T_2 is isomorphic to the Markov shift with transition matrix $\begin{bmatrix} q & p \\ p & q \end{bmatrix}$.

Since $T_2^{-n}\mathcal{B} = T_1^{-n}\mathcal{B}$ $(n \geq 0)$ we know that T_2 is also exact. We have $T_1^2 = T_2^2$ (which implies $h(T_1) = h(T_2)$), $T_1^{-n}\mathcal{B} = T_2^{-n}\mathcal{B}$ $(n \geq 0)$, $j_{T_1}(x) = j_{T_2}(x)$ and $\beta(T_1) = \beta(T_2)$. However $T_1 \not\cong T_2$. To see this suppose $T_1 \overset{\phi}{\cong} T_2$. Then $\phi\beta(T_1) = \beta(T_2)$ and by theorem 2 ϕ has the form $\phi(y,g) = (y, \lambda(y,g))$. Then $F(y,g) = \lambda(y,g)\cdot g^{-1}$ satisfies $F(T_1) = -F$, which contradicts the weak-mixing of T_1.

After doing this work we found that Vinokurov [6] had also constructed the above example.

A similar example could be obtained as follows. Again let $S : (Y,\mathcal{C},\mu) \to Y$ be the 1-sided (p,q) shift $(p \neq q)$. Let G be a compact monothetic group (multiplicative notation) equipped with Haar measure and put $X = Y \times G$. Choose a measurable map $\alpha : Y \to G$ so that $T_1(y,g) = (Sy, \alpha(y)\cdot g)$ is weak-mixing. (Such an α can always be chosen [2]). It follows that T_1 is exact. Suppose G has an element $g_0 \neq e$ of finite order $(g_0^k = e)$, and define T_2 by $T_2(y,g) = (Sy, g_0\, \alpha(y) g)$. Then $T_2^k = T_1^k$, $T_1^{-n}\mathcal{B} = T_2^{-n}\mathcal{B}$ $(n \geq 0)$, $j_{T_1} = j_{T_2}$, $\beta(T_1) = \beta(T_2)$, but $T_1 \not\cong T_2$.

The reason that T_1 and T_2 are not isomorphic is that they have different "eigenvalues mod $\beta(T)$". We shall explain this in the next section. It turns out that every ergodic endomorphism with discrete spectrum mod $\beta(T)$ is isomorphic to an endomorphism of the form $(y,g) \to (Sy, \alpha(y)\cdot g)$ $y \in Y$, $g \in G$ where $S : (Y, \mathcal{C}, \mu) \to Y$ is an endomorphism with $\beta(S) = \mathcal{C}$, G is a compact abelian group, and $\alpha : Y \to G$ is measurable.

(4) Discrete spectrum mod $\beta(T)$

Let $T : X \to X$ be an endomorphism and $\mathcal{Q}$ a sub σ-algebra of $\mathcal{B}$ with $T^{-1}\mathcal{Q} \subset \mathcal{Q}$. $T : X \to X$ has discrete spectrum mod $\mathcal{Q}$ if $\mathcal{G} = \mathcal{G}_T(\mathcal{Q}) = \{f \in L^2(\mathcal{B}) \mid |f| = 1, \frac{f(T)}{f} \in L^2(\mathcal{Q})\}$ spans $L^2(\mathcal{B})$. The classical discrete spectrum case corresponds to $\mathcal{Q} = \mathcal{N}$. $\mathcal{G}$ is a group under pointwise multiplication and is called the group of eigenfunctions of T mod $\mathcal{Q}$.

THEOREM 4 (Representation theorem)

If $T : X \to X$ is ergodic and has discrete spectrum mod $\mathcal{Q}$ ($T^{-1}\mathcal{Q} \subset \mathcal{Q}$) then X is isomorphic to $X_{\mathcal{Q}} \times G$ where $X_{\mathcal{Q}}$ is the quotient space of X by $\mathcal{Q}$ and G is a compact abelian group. In this representation T has the form $T(y,g) = (Sy, \alpha(y)\cdot g)$ where $T_{\mathcal{Q}} = S : X_{\mathcal{Q}} \to X_{\mathcal{Q}}$ is the endomorphism induced by T on $X_{\mathcal{Q}}$ and $\alpha : Y \to G$ is measurable. G is the dual of the countable group $\mathcal{G}/\{f \in L^2(\mathcal{Q}) \mid |f| = 1\}$:

This result is a generalisation of the theorem that represents ergodic transformations with discrete spectrum as rotations on compact abelian groups. It corresponds to the case $\mathcal{Q} = \mathcal{N}$.

$\mathcal{H} = \mathcal{H}_T(\mathcal{Q}) = \{ \frac{f(T)}{f} \mid f \in \mathcal{G}\}$ is called the group of eigenvalues of T mod $\mathcal{Q}$. When T is ergodic $\mathcal{G}/\{\frac{f(T)}{f} \mid f \in L^2(\mathcal{Q}), |f| = 1\}$ is a countable group isomorphic to $\mathcal{G}/\{f \mid f \in L^2(\mathcal{Q}), |f| = 1\}$. When T is ergodic and has the canonical form $T(y,g) = (Sy, \alpha(y)g)$ given by theorem 4 then $\mathcal{G} = \{h\cdot\gamma \mid h \in L^2(\mathcal{Q}), |h| = 1, \gamma \in \hat{G}\}$ and $\mathcal{H} = \{\frac{h(Sy)}{h(y)} \gamma(\alpha(y)) \mid h. \gamma \in \mathcal{G}\}$. We state now the generalisations of the existence and isomorphism theorems for automorphisms with discrete spectrum to the case of endomorphisms with discrete spectrum mod $\beta(T)$.

THEOREM 5 (EXISTENCE THEOREM)

Let $S : (Y, \mathcal{C}, \mu) \to Y$ be an ergodic endomorphism with $\beta(S) = \mathcal{C}$ and let $\mathcal{H}$ be a subgroup of $\{f \mid f \in L^2(\mathcal{C}), |f| = 1\}$ so that $\mathcal{H} \supset \{\frac{f(S)}{f} \mid f \in L^2(\mathcal{C}), |f| = 1\}$ and $\mathcal{H}/\{\frac{f(S)}{f} \mid f \in L^2(\mathcal{C}), |f| = 1\}$ is countable. If G is the compact abelian group with $\hat{G} = \mathcal{H}/\{\frac{f(S)}{f} \mid f \in L^2(\mathcal{C}), |f| = 1\}$ there is an ergodic endomorphism T of $X = Y \times G$ of the form $T(y,g) = (Sy, \alpha(y)g)$ with $\beta(T) = \pi^{-1}(\mathcal{C})$, T has discrete spectrum mod $\beta(T)$ and the group of eigenvalues of T mod $\beta(T)$ is $\{h \circ \pi \mid h \in \mathcal{H}\}$.

THEOREM 6 (ISOMORPHISM THEOREM)

Let $T_1, T_2 : X \to X$ be ergodic, $\beta(T_1) = \beta(T_2)$, $T_{\beta(T_1)} = T_{\beta(T_2)}$ and let both have discrete spectrum mod $\beta(T_1)$. Then $T_1 \cong T_2 \Leftrightarrow$ they have identical groups of eigenvalues mod $\beta(T_1)$.

Consider now the canonical set-up where $S : (Y, \mathcal{C}, \mu) \to Y$ is an endomorphism with $\beta(S) = \mathcal{C}$, G is a compact abelian group, $X = Y \times G$, and $T(y,g) = (Sy, \alpha(y)g)$ where $\alpha : Y \to G$ is measurable. For each measurable $\delta : Y \to G$ define T_δ by $T_\delta(y,g) = (Sy, \delta(y)\,\alpha(y)g)$. If T and T_δ are ergodic then $T \cong T_\delta$ implies that $\delta(y) = \frac{h(T(y,g))}{h(y,g)}$ can be solved for a measurable map $h : X \to G$ (i.e. δ is a coboundary for T). If we put $\delta(y) = \text{constant} = g_0$, then if T, T_{g_0} are ergodic and isomorphic we can conclude that $\gamma(g_0)$ is an eigenvalue of T for each $\gamma \in \hat{G}$. So if T is weak-mixing all the transformations T_g, $g \in G$ are non-isomorphic. This gives an uncountable collection (if G is uncountable) of exact endomorphisms with equal entropy, identical jacobians, identical $\beta(T)$-algebras and identical sequences $\{T^{-n}\mathcal{B}\}_0^\infty$, but no two of them are isomorphic.

REFERENCES

[1] H. Furstenberg — Disjointness in Ergodic Theory. Math. Systems Theory 1 (1967) 1-50.

[2] R. Jones and W. Parry — Compact abelian group extensions of dynamical systems. Compositio Math.

[3] W. Parry — Entropy and generations in Ergodic Theory Benjamin 1969.

[4] W. Parry and P. Walters — Endomorphisms of a Lebesque space Bull A. M. S. 78 1972 272-276.

[5] V. A. Rohlin — On the fundamental ideas of measure theory. A. M. S. translations Series 1. vol. 10 p.1-54.

[6] V. G. Vinokurov — Two nonisomorphic exact endomorphisms of a Lebesque space with isomorphic decomposition sequences. "Fan" Uzbek S.S.R. Tashlcent 1970 43-45.

[7] P. Walters — On roots of n : 1 measure-preserving transformations. Journal London Math. Soc. 44(1969), 7-14.

The University of Warwick
Coventry
England.

GROUPS OF MEASURE PRESERVING TRANSFORMATIONS

BENJAMIN WEISS

HEBREW UNIVERSITY

An important object in statistical mechanics is a discrete random field, which may be described as follows. At each lattice point, $\ell \in \mathbb{Z}^d$ a state is assigned from some finite set $S = \{1,2, \ldots s\}$ in a random fashion. More precisely one is given a measure on the space $X = S^{(\mathbb{Z}^d)}$, where each $x \in X$ represents a possible global state of the system. In many cases it is natural to assume that the measure μ on X is translation invariant, and if we represent the effect of translation in $\mathbb{Z}^d$ by ℓ on points of X by the mapping $\varphi^\ell : X \longrightarrow X$ then we have a group $\Phi = \{\varphi^\ell\}$ of measure preserving transformations. In this way statistical mechanics leads us in a natural fashion to a multiparametered version of ergodic theory, and I'll try now to describe briefly some recent results concerning such groups.

A basic tool is a theorem first adumbrated by S. Kakutani and then given explicitly by V. A. Rokhlin [7] in the case $d = 1$, which in our setting reads as follows. A group is said to be non-periodic if for all $\ell \neq 0$, $\mu\{x: \varphi^\ell x = x\} = 0$.

Theorem 1: If Φ is non-periodic, then given $0 < \ell \in Z^d$, and $\epsilon > 0$, there is a set $E \subset X$ such that (i) $\{\varphi^{\ell'} E\}$ are disjoint for $0 \le \ell' \le \ell$

(ii) $\mu(\bigcup_{0 \le \ell' \le \ell} \varphi^{\ell'} E) > 1 - \varepsilon$

(iii) $\mu(E \,\Delta\, \varphi^{\ell_i + 1} E) < \varepsilon$, $1 \le i \le d$

Furthermore, if we are given a finite partition $\alpha = \{A_1, A_2, \ldots, A_k\}$

then E may be chosen to satisfy (i)-(iii) and in addition

$$\text{(iv)} \qquad \frac{\mu(A_j \cap \varphi^{\ell'} E)}{\mu(\varphi^{\ell} E)} = \mu(A_j), \quad 1 \leq j \leq k, \quad 0 \leq \ell' \leq \ell.$$

For $d = 1$, (iii) represents no improvement, but (iv) is a sharpening of the classical result even for $d = 1$. Several authors have discussed extending entropy theory to groups Φ, see [3] and [8] for example. The entropy $h(\Phi)$ is of course an <u>isomorphism</u> invariant, where $\Phi \cong \tilde{\Phi}$ means that there is a one to one mapping $\vartheta : X \longrightarrow \tilde{X}$ such that $\tilde{\mu}(\vartheta E) = \mu(E)$ and for all $\varphi^{\ell} \in \Phi$, $\tilde{\varphi}^{\ell} \in \tilde{\Phi}$ we have $\tilde{\varphi}^{\ell} \vartheta = \vartheta \varphi^{\ell}$. In case the random field described at the beginning has no interaction then we have at each point of Z^d an independent random invariable and μ becomes simply product measure on S^{Z^d} of a distribution $(p_1, p_2, \cdots p_s)$. As when $d = 1$, it turns out that for these measures $h(\Phi) = - \Sigma\, p_i \log p_i$, and we call such groups <u>Bernoullian</u>. After getting a Shannon-McMillan theorem for the entropy of ergodic groups Φ, one can use theorem 1 and the methods developed by D. Ornstein [4], to prove:

<u>Theorem 2</u>: Two Bernoullian groups with the same entropy are isomorphic.

This settles a question raised in [12], where a fuller discussion may be found of the isomorphism problem.

From the investigations of Dobrushin, Spitzer, Sherman and others (see [9] and [10]) one knows that the Gibbsian random fields are precisely those that satisfy a strong conditional independence requirement called <u>Markov random field</u> (MRF). For various notions of approximate independence it is possible to extend theorem 2 as was done for the one dimensional case in [5], [6]. It seems very likely that MRF implies one of these notions so that we would get the result that fields with interactions are isomorphic to fields where the various positions are independent. Details of these results can be found in a joint paper with Yitzhak Katznelson [2]. Similar results were obtained independently by J. P. Conze [1] and J.P. J.P. Thouvenot [11].

References

1. J.P. Conze, Entropie d'un Group Abelien de Transformations. (preprint).

2. Y. Katznelson and B. Weiss, Commuting Measure Preserving Transformations, Israel Journal of Mathematics (to appear).

3. A.A. Kirillov, Dynamical Systems, Factors and Group Representations Usp, Mat. Nauk 22 (1967), 67-80.

4. D.S. Ornstein, Bernoulli Shifts with the same Entropy are Isomorphic, Adv. in Math. 4(1970), 337-352.

5. __________, Two Bernoulli Shifts with Infinite Entropy are Isomorphic, ibid. 5 (1970), 339-348.

6. __________, Factors of Bernoulli Shifts are Bernoulli Shifts, ibid. 5(1970), 349-364.

7. V.A. Rokhlin, A General Measure Preserving Transformation is not mixing. Doklady Abad. Nauk. SSU v. 60 (1948), 349-51.

8. D. Ruelle, Statistical Mechanics, N.Y., 1969.

9. S. Sherman, Markov Random Fields and Gibbs Random Fields. Israel Journal of Math., (to appear).

10. F. Spitzer, Markov Random Fields and Gibbs Ensembles, Amer. Math. Monthly 79(1971), 142-154.

11. J.P. Thouvenot, Convergence en Moyenne de l'Information pour l'action de $\mathbb{Z}^2$. (preprint).

12. B. Weiss, The Isomorphism Problem in Ergodic Theory. BAMS (to appear).

CLASSIFICATION OF SUBSHIFTS OF FINITE TYPE

R. F. Williams

Northwestern University

The reader is referred to the bibliography, especially [1] and [7], for any discussion of the history of symbolic dynamics. Suffice it to say that the definition of subshift of finite type is due to Parry [3,4] and independently evolved from Smale [5] via Bowen-Lanford [2]. See [7] for proofs and related results.

Let A be an n×n 0-1 matrix. A two letter word ij is _allowable_ provided $A_{ij} = 1$. The corresponding _subshift of finite type_ $\hat{a}:S(A) \to S(A)$ is defined as follows:

1) S(A) is the set of all doubly infinite sequences $x = (x_i)_{i \in \mathbb{Z}}$ such that

   i) $x_i \in \{1,\ldots,n\}$; and

   ii) $x_i x_{i+1}$ is allowable for all $i \in \mathbb{Z}$.

2) $[\hat{a}(x)]_i = x_{i+1}$.

S(A) is a compact, 0-dimensional metric space with (say) the metric

$$d(x,y) = \sum_{i=-\infty}^{\infty} \eta_{ij} 2^{-|i|},$$

where $\eta_{ij} = 1$ if $i \neq j$, 0 otherwise. The one-sided shift $a:C(A) \to C(A)$ is defined similarly except that coordinates x_i occur only for $i \geq 0$. $\hat{a}$ is a homeomorphism; a is finite-to-one and open.

Example. Let

$$SR = \begin{pmatrix} 1 & 1 & 0 \\ 0 & 0 & 1 \\ 0 & 0 & 1 \end{pmatrix} \begin{pmatrix} 1 & 0 & 0 \\ 0 & 1 & 0 \\ 1 & 0 & 1 \end{pmatrix} = \begin{pmatrix} 1 & 1 & 0 \\ 1 & 0 & 1 \\ 1 & 0 & 1 \end{pmatrix} = A$$

$$RS = \begin{pmatrix} 1 & 0 & 0 \\ 0 & 1 & 0 \\ 1 & 0 & 1 \end{pmatrix} \begin{pmatrix} 1 & 1 & 0 \\ 0 & 0 & 1 \\ 0 & 0 & 1 \end{pmatrix} = \begin{pmatrix} 1 & 1 & 0 \\ 0 & 0 & 1 \\ 1 & 1 & 1 \end{pmatrix} = B.$$

Then the allowable 2-letter words are

aa, ab, ac, ba, cb, cc for S(A)

$\alpha\alpha$, $\alpha\gamma$, $\beta\alpha$, $\beta\gamma$, $\gamma\beta$, $\gamma\gamma$ for S(B)

where we have used Latin for S(A), Greek for S(B).

The factorization SR = A allows us to define _allowable 3-letter words_ ijk whenever ij is allowable because

$$1 = A_{ik} = \sum_{\ell} S_{i\ell}R_{\ell k} = S_{ij}R_{jk} \quad \text{for a unique } j.$$

Thus the allowable 3-letter words (interpreting the j as Greek) are:

$$a\alpha a,\ a\gamma b,\ a\gamma c,\ b\beta a,\ c\gamma b,\ c\gamma c.$$

Hence R induces a map $\hat{r}$: $S(A) \to S(B)$ where

$$[\hat{r}(x)]_i = \begin{cases} \alpha \text{ if } x_i x_{i+1} = aa \\ \beta \text{ if } x_i x_{i+1} = ba \\ \gamma \text{ if } x_i x_{i+1} = ab, ac, cb \text{ or } cc. \end{cases}$$

Now suppose $x \in S(A)$ and its 3-coordinates beginning with the ith are $jk\ell$, and those of $\hat{r}(x)$ are uvw. Then

$$A_{jk}A_{k\ell} = 1 = S_{ju}R_{uk}S_{kv}R_{v\ell}$$

so that

$$R_{uk}S_{kv} = 1.$$

Thus the ith coordinate of $\hat{s}\hat{r}(x)$ is k. That is, $\hat{s}\hat{r}$ shifts the coordinates of S(A) by 1 or, $\hat{s}\hat{r} = \hat{a}$. Similarly $\hat{r}\hat{s} = \hat{b}$. In particular, $\hat{r}$ is a homeomorphism.

Theorem. Any conjugacy $h: S(A) \to S(B)$ between two shifts of finite type is the composition of finitely many conjugacies, defined just as in our example.

The proof occupies the bulk of the paper [7]; in particular, no algorithm is given for determining the various stages. The point, rather, is that these techniques give a classification of subshifts of finite type.

The functoral nature of these induced maps ($\hat{r}\hat{s} = \hat{a}$ because RS = A) is worthy of exploitation. This points out a weakness of 0-1 matrices: they are not closed under products! In [7] S(A) and C(A) are defined for arbitrary n×n matrices over $\mathbb{Z}^+$. This is not hard, but we don't repeat it here.

Definition. First we formalize the relations occurring in Example 1. For R, S rectangular over $\mathbb{Z}^+$, A = RS, B = SR we say $A \sim_e B$. Making $\sim_e$ transitive, by allowing finite strings, we obtain an equivalence relation, $\sim_S$, called strong shift equivalence. Example 1 shows

Theorem. If two matrices are strongly shift equivalent, then the corresponding subshifts are topologically conjugate.

This is easy, but very preliminary as $\sim_S$ is so non-computable; a priori an equivalence between two 2×2 matrices could involve an intermediate matrix of order 50. A consequence of $\sim_S$ is <u>shift equivalence</u> ($\sim_s$): $A \sim_s B$ provided there are matrices R, S and an integer m such that

$$RA = BR, \quad AS = SB, \quad SR = A^m, \quad RS = B^m.$$

m is called the <u>lag</u> of the <u>equivalence</u>, R, S. One checks that R, S is an equivalence of lag 1 from SR to RS, so that $\sim_S \Rightarrow \sim_s$.

In many "categories" the other implication ($\sim_s \Rightarrow \sim_S$) is easy, as illustrated here in the lag 2 case: Given

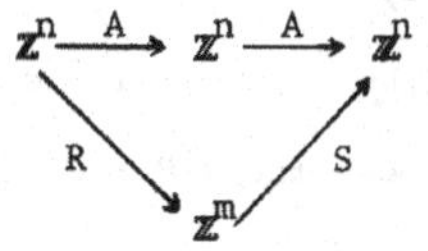

we "triangulate" it to get

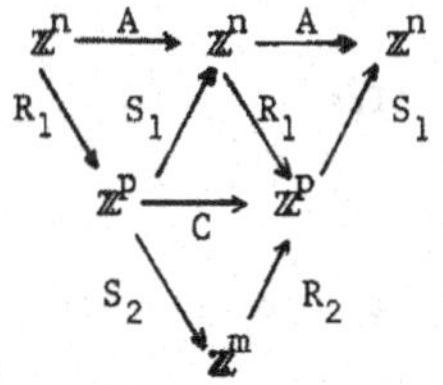

The trouble is, matrices over $\mathbb{Z}^+$ do not occur as the maps in any ordinary "category." However, a roundabout proof is given in [7], so that

<u>Theorem</u>. Two subshifts of finite type are topologically conjugate iff the corresponding matrices are shift equivalent.

Actually our principal tool (or "category") is 1-sided subshifts of finite type and "Markov" maps between them. Almost as a byproduct, we obtain a simple and effective classification of these 1-sided shifts, as follows:

<u>Definitions</u>. A matrix S is a subdivision matrix provided each row has only one non-zero term and this term is a 1. Say A is a reduction of B ($A < B$) provided there are R, S over $\mathbb{Z}^+$ with S a subdivision matrix such that $A = RS$, $B = SR$.

There are two cases — either S is square 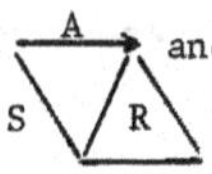 and hence a permutation matrix, or S and (therefore) B have repeated rows.

Lemma. Each matrix A has a complete reduction, A_0. A reduction of A is complete iff it has no repeated rows.

Theorem. If A and B are square matrices over $\mathbb{Z}^+$ then the corresponding 1-sided shifts are topologically conjugate iff their complete reductions A_0, B_0 are the same up to a permutation of the basis elements.

Example.

$$A = \begin{pmatrix} 1 & 1 & 0 \\ 1 & 0 & 1 \\ 1 & 0 & 1 \end{pmatrix} > \begin{pmatrix} 1 & 1 \\ 1 & 1 \end{pmatrix} > (2) = A_0,$$

which is completely reduced. Algorithm: If rows i and j of A are identical, then add the jth column to the ith and remove the jth row and column for the result, to get a reduction A_1 of A.

Example. (3) and $\begin{pmatrix} 1 & 3 & 1 \\ 1 & 1 & 0 \\ 0 & 2 & 1 \end{pmatrix}$ represent the full 2-sided shift on three symbols, but non-conjugate 1-sided shifts.

Three advantages of matrices over $\mathbb{Z}^+$:

1) Closed under products. E.g., any power $\hat{a}^j$ of a shift $\hat{a}$ is a subshift of finite type. Its matrix is A^j.

2) Many examples are available with 2×2 matrices and hence computable. Thus $\begin{pmatrix} 4 & 1 \\ 1 & 0 \end{pmatrix}$ and $\begin{pmatrix} 3 & 2 \\ 2 & 1 \end{pmatrix}$ determine non-conjugate shifts, though they have the same zeta function, i.e., the same number of periodic points for all periods.

3) In this setting, there is a canonical matrix for 1-sided shifts.

Problems.

1. Classify endomorphisms of 1- and 2-sided shifts of finite type. Our methods give some information only when the endomorphism is open and finite-to-one.

2. If S(A) has k fixed points under $\hat{a}$, is there an automorphism of S(A), commuting with $\hat{a}$ and realizing any permutations of the fixed points?

3. Is there an effective algorithm for deciding when two matrices over $\mathbb{Z}^+$ are shift equivalent? For 2×2 matrices of determinant I?

Bibliography

[1] R. Bowen, Markov partitions for axiom A diffeomorphisms, Amer. J. Math. 42 (1970), 725-747.

[2] ——— and O. Lanford, Zeta functions of restrictions of the shift transformation, ibid, pp. 43-50.

[3] W. Parry, Symbolic dynamics and transformations of the unit interval, Trans. Amer. Math. Soc. 122 (1966), 368-378.

[4] ———, Intrinsic Markov chains, Trans. Amer. Math. Soc. 112 (1964), 55-66.

[5] S. Smale, Diffeomorphisms with many periodic points, Differential and Combinatorial Topology (A symposium in honor of Marston Morse), Princeton University Press, Princeton, N. J., pp. 63-80.

[6] R. Williams, Classification of symbol spaces of finite type, Bull. Amer. Math. Soc. 77 (1971), 439-443.

[7] ———, Classification of subshifts of finite type, to appear. Preprint, Northwestern University.

Bibliography

(1) [illegible]

(2) [illegible]

(3) [illegible]

(4) [illegible]

(5) [illegible]

(6) [illegible]

Lecture Notes in Mathematics

Comprehensive leaflet on request

Vol. 146: A. B. Altman and S. Kleiman, Introduction to Grothendieck Duality Theory. II, 192 pages. 1970. DM 18,–

Vol. 147: D. E. Dobbs, Cech Cohomological Dimensions for Commutative Rings. VI, 176 pages. 1970. DM 16,–

Vol. 148: R. Azencott, Espaces de Poisson des Groupes Localement Compacts. IX, 141 pages. 1970. DM 16,–

Vol. 149: R. G. Swan and E. G. Evans, K-Theory of Finite Groups and Orders. IV, 237 pages. 1970. DM 20,–

Vol. 150: Heyer, Dualität lokalkompakter Gruppen. XIII, 372 Seiten. 1970. DM 20,–

Vol. 151: M. Demazure et A. Grothendieck, Schémas en Groupes I. (SGA 3). XV, 562 pages. 1970. DM 24,–

Vol. 152: M. Demazure et A. Grothendieck, Schémas en Groupes II. (SGA 3). IX, 654 pages. 1970. DM 24,–

Vol. 153: M. Demazure et A. Grothendieck, Schémas en Groupes III. (SGA 3). VIII, 529 pages. 1970. DM 24,–

Vol. 154: A. Lascoux et M. Berger, Variétés Kähleriennes Compactes. VII, 83 pages. 1970. DM 16,–

Vol. 155: Several Complex Variables I, Maryland 1970. Edited by J. Horváth. IV, 214 pages. 1970. DM 18,–

Vol. 156: R. Hartshorne, Ample Subvarieties of Algebraic Varieties. XIV, 256 pages. 1970. DM 20,–

Vol. 157: T. tom Dieck, K. H. Kamps und D. Puppe, Homotopietheorie. VI, 265 Seiten. 1970. DM 20,–

Vol. 158: T. G. Ostrom, Finite Translation Planes. IV. 112 pages. 1970. DM 16,–

Vol. 159: R. Ansorge und R. Hass. Konvergenz von Differenzenverfahren für lineare und nichtlineare Anfangswertaufgaben. VIII, 145 Seiten. 1970. DM 16,–

Vol. 160: L. Sucheston, Constributions to Ergodic Theory and Probability. VII, 277 pages. 1970. DM 20,–

Vol. 161: J. Stasheff, H-Spaces from a Homotopy Point of View. VI, 95 pages. 1970. DM 16,–

Vol. 162: Harish-Chandra and van Dijk, Harmonic Analysis on Reductive p-adic Groups. IV, 125 pages. 1970. DM 16,–

Vol. 163: P. Deligne, Equations Différentielles à Points Singuliers Reguliers. III, 133 pages. 1970. DM 16,–

Vol. 164: J. P. Ferrier, Seminaire sur les Algebres Complètes. II, 69 pages. 1970. DM 16,–

Vol. 165: J. M. Cohen, Stable Homotopy. V, 194 pages. 1970. DM 16,–

Vol. 166: A. J. Silberger, PGL_2 over the p-adics: its Representations, Spherical Functions, and Fourier Analysis. VII, 202 pages. 1970. DM 18,–

Vol. 167: Lavrentiev, Romanov and Vasiliev, Multidimensional Inverse Problems for Differential Equations. V, 59 pages. 1970. DM 16,–

Vol. 168: F. P. Peterson, The Steenrod Algebra and its Applications: A conference to Celebrate N. E. Steenrod's Sixtieth Birthday. VII, 317 pages. 1970. DM 22,–

Vol. 169: M. Raynaud, Anneaux Locaux Henséliens. V, 129 pages. 1970. DM 16,–

Vol. 170: Lectures in Modern Analysis and Applications III. Edited by C. T. Taam. VI, 213 pages. 1970. DM 18,–

Vol. 171: Set-Valued Mappings, Selections and Topological Properties of 2^X. Edited by W. M. Fleischman. X, 110 pages. 1970. DM 16,–

Vol. 172: Y.-T. Siu and G. Trautmann, Gap-Sheaves and Extension of Coherent Analytic Subsheaves. V, 172 pages. 1971. DM 16,–

Vol. 173: J. N. Mordeson and B. Vinograde, Structure of Arbitrary Purely Inseparable Extension Fields. IV, 138 pages. 1970. DM 16,–

Vol. 174: B. Iversen, Linear Determinants with Applications to the Picard Scheme of a Family of Algebraic Curves. VI, 69 pages. 1970. DM 16,–

Vol. 175: M. Brelot, On Topologies and Boundaries in Potential Theory. VI, 176 pages. 1971. DM 18,–

Vol. 176: H. Popp, Fundamentalgruppen algebraischer Mannigfaltigkeiten. IV, 154 Seiten. 1970. DM 16,–

Vol. 177: J. Lambek, Torsion Theories, Additive Semantics and Rings of Quotients. VI, 94 pages. 1971. DM 16,–

Vol. 178: Th. Bröcker und T. tom Dieck, Kobordismentheorie. XVI, 191 Seiten. 1970. DM 18,–

Vol. 179: Seminaire Bourbaki – vol. 1968/69. Exposés 347-363. IV. 295 pages. 1971. DM 22,–

Vol. 180: Séminaire Bourbaki – vol. 1969/70. Exposés 364-381. IV, 310 pages. 1971. DM 22,–

Vol. 181: F. DeMeyer and E. Ingraham, Separable Algebras over Commutative Rings. V, 157 pages. 1971. DM 16.–

Vol. 182: L. D. Baumert. Cyclic Difference Sets. VI, 166 pages. 1971. DM 16,–

Vol. 183: Analytic Theory of Differential Equations. Edited by P. F. Hsieh and A. W. J. Stoddart. VI, 225 pages. 1971. DM 20,–

Vol. 184: Symposium on Several Complex Variables, Park City, Utah, 1970. Edited by R. M. Brooks. V, 234 pages. 1971. DM 20,–

Vol. 185: Several Complex Variables II, Maryland 1970. Edited by J. Horváth. III, 287 pages. 1971. DM 24,–

Vol. 186: Recent Trends in Graph Theory. Edited by M. Capobianco/ J. B. Frechen/M. Krolik. VI, 219 pages. 1971. DM 18.–

Vol. 187: H. S. Shapiro, Topics in Approximation Theory. VIII, 275 pages. 1971. DM 22,–

Vol. 188: Symposium on Semantics of Algorithmic Languages. Edited by E. Engeler. VI, 372 pages. 1971. DM 26,–

Vol. 189: A. Weil, Dirichlet Series and Automorphic Forms. V, 164 pages. 1971. DM 16,–

Vol. 190: Martingales. A Report on a Meeting at Oberwolfach, May 17-23, 1970. Edited by H. Dinges. V, 75 pages. 1971. DM 16,–

Vol. 191: Séminaire de Probabilités V. Edited by P. A. Meyer. IV, 372 pages. 1971. DM 26,–

Vol. 192: Proceedings of Liverpool Singularities – Symposium I. Edited by C. T. C. Wall. V, 319 pages. 1971. DM 24,–

Vol. 193: Symposium on the Theory of Numerical Analysis. Edited by J. Ll. Morris. VI, 152 pages. 1971. DM 16,–

Vol. 194: M. Berger, P. Gauduchon et E. Mazet. Le Spectre d'une Variété Riemannienne. VII, 251 pages. 1971. DM 22,–

Vol. 195: Reports of the Midwest Category Seminar V. Edited by J.W. Gray and S. Mac Lane. III, 255 pages. 1971. DM 22,–

Vol. 196: H-spaces – Neuchâtel (Suisse)- Août 1970. Edited by F. Sigrist, V, 156 pages. 1971. DM 16,–

Vol. 197: Manifolds – Amsterdam 1970. Edited by N. H. Kuiper. V, 231 pages. 1971. DM 20,–

Vol. 198: M. Hervé, Analytic and Plurisubharmonic Functions in Finite and Infinite Dimensional Spaces. VI, 90 pages. 1971. DM 16.–

Vol. 199: Ch. J. Mozzochi, On the Pointwise Convergence of Fourier Series. VII, 87 pages. 1971. DM 16,–

Vol. 200: U. Neri, Singular Integrals. VII, 272 pages. 1971. DM 22,–

Vol. 201: J. H. van Lint, Coding Theory. VII, 136 pages. 1971. DM 16,–

Vol. 202: J. Benedetto, Harmonic Analysis on Totally Disconnected Sets. VIII, 261 pages. 1971. DM 22,–

Vol. 203: D. Knutson, Algebraic Spaces. VI, 261 pages. 1971. DM 22,–

Vol. 204: A. Zygmund, Intégrales Singulières. IV, 53 pages. 1971. DM 16,–

Vol. 205: Séminaire Pierre Lelong (Analyse) Année 1970. VI, 243 pages. 1971. DM 20,–

Vol. 206: Symposium on Differential Equations and Dynamical Systems. Edited by D. Chillingworth. XI, 173 pages. 1971. DM 16,–

Vol. 207: L. Bernstein, The Jacobi-Perron Algorithm – Its Theory and Application. IV, 161 pages. 1971. DM 16,–

Vol. 208: A. Grothendieck and J. P. Murre, The Tame Fundamental Group of a Formal Neighbourhood of a Divisor with Normal Crossings on a Scheme. VIII, 133 pages. 1971. DM 16,–

Vol. 209: Proceedings of Liverpool Singularities Symposium II. Edited by C. T. C. Wall. V, 280 pages. 1971. DM 22,–

Vol. 210: M. Eichler, Projective Varieties and Modular Forms. III, 118 pages. 1971. DM 16,–

Vol. 211: Théorie des Matroïdes. Edité par C. P. Bruter. III, 108 pages. 1971. DM 16,–

Please turn over

Vol. 212: B. Scarpellini, Proof Theory and Intuitionistic Systems. VII, 291 pages. 1971. DM 24,–

Vol. 213: H. Hogbe-Nlend, Théorie des Bornologies et Applications. V, 168 pages. 1971. DM 18,–

Vol. 214: M. Smorodinsky, Ergodic Theory, Entropy. V, 64 pages. 1971. DM 16,–

Vol. 215: P. Antonelli, D. Burghelea and P. J. Kahn, The Concordance-Homotopy Groups of Geometric Automorphism Groups. X, 140 pages. 1971. DM 16,–

Vol. 216: H. Maaß, Siegel's Modular Forms and Dirichlet Series. VII, 328 pages. 1971. DM 20,–

Vol. 217: T. J. Jech, Lectures in Set Theory with Particular Emphasis on the Method of Forcing. V, 137 pages. 1971. DM 16,–

Vol. 218: C. P. Schnorr, Zufälligkeit und Wahrscheinlichkeit. IV, 212 Seiten 1971. DM 20,–

Vol. 219: N. L. Alling and N. Greenleaf, Foundations of the Theory of Klein Surfaces. IX, 117 pages. 1971. DM 16,–

Vol. 220: W. A. Coppel, Disconjugacy. V, 148 pages. 1971. DM 16,–

Vol. 221: P. Gabriel und F. Ulmer, Lokal präsentierbare Kategorien. V, 200 Seiten. 1971. DM 18,–

Vol. 222: C. Meghea, Compactification des Espaces Harmoniques. III, 108 pages. 1971. DM 16,–

Vol. 223: U. Felgner, Models of ZF-Set Theory. VI, 173 pages. 1971. DM 16,–

Vol. 224: Revètements Etales et Groupe Fondamental. (SGA 1). Dirigé par A. Grothendieck XXII, 447 pages. 1971. DM 30,–

Vol. 225: Théorie des Intersections et Théorème de Riemann-Roch. (SGA 6). Dirigé par P. Berthelot, A. Grothendieck et L. Illusie. XII, 700 pages. 1971. DM 40,–

Vol. 226: Seminar on Potential Theory, II. Edited by H. Bauer. IV, 170 pages. 1971. DM 18,–

Vol. 227: H. L. Montgomery, Topics in Multiplicative Number Theory. IX, 178 pages. 1971. DM 18,–

Vol. 228: Conference on Applications of Numerical Analysis. Edited by J. Ll. Morris. X, 358 pages. 1971. DM 26,–

Vol. 229: J. Väisälä, Lectures on n-Dimensional Quasiconformal Mappings. XIV, 144 pages. 1971. DM 16,–

Vol. 230: L. Waelbroeck, Topological Vector Spaces and Algebras. VII, 158 pages. 1971. DM 16,–

Vol. 231: H. Reiter, L^1-Algebras and Segal Algebras. XI, 113 pages. 1971. DM 16,–

Vol. 232: T. H. Ganelius, Tauberian Remainder Theorems. VI, 75 pages. 1971. DM 16,–

Vol. 233: C. P. Tsokos and W. J. Padgett. Random Integral Equations with Applications to Stochastic Systems. VII, 174 pages. 1971. DM 18,–

Vol. 234: A. Andreotti and W. Stoll. Analytic and Algebraic Dependence of Meromorphic Functions. III, 390 pages. 1971. DM 26,–

Vol. 235: Global Differentiable Dynamics. Edited by O. Hájek, A. J. Lohwater, and R. McCann. X, 140 pages. 1971. DM 16,–

Vol. 236: M. Barr, P. A. Grillet, and D. H. van Osdol. Exact Categories and Categories of Sheaves. VII, 239 pages. 1971, DM 20,–

Vol. 237: B. Stenström. Rings and Modules of Quotients. VII, 136 pages. 1971. DM 16,–

Vol. 238: Der kanonische Modul eines Cohen-Macaulay-Rings. Herausgegeben von Jürgen Herzog und Ernst Kunz. VI, 103 Seiten. 1971. DM 16,–

Vol. 239: L. Illusie, Complexe Cotangent et Déformations I. XV, 355 pages. 1971. DM 26,–

Vol. 240: A. Kerber, Representations of Permutation Groups I. VII, 192 pages. 1971. DM 18,–

Vol. 241: S. Kaneyuki, Homogeneous Bounded Domains and Siegel Domains. V, 89 pages. 1971. DM 16,–

Vol. 242: R. R. Coifman et G. Weiss, Analyse Harmonique Non-Commutative sur Certains Espaces. V, 160 pages. 1971. DM 16,–

Vol. 243: Japan-United States Seminar on Ordinary Differential and Functional Equations. Edited by M. Urabe. VIII, 332 pages. 1971. DM 26,–

Vol. 244: Séminaire Bourbaki – vol. 1970/71. Exposés 382–399. IV, 356 pages. 1971. DM 26,–

Vol. 245: D. E. Cohen, Groups of Cohomological Dimension One. V, 99 pages. 1972. DM 16,–

Vol. 246: Lectures on Rings and Modules. Tulane University Ring and Operator Theory Year, 1970–1971. Volume I. X, 661 pages. 1972. DM 40,–

Vol. 247: Lectures on Operator Algebras. Tulane University Ring and Operator Theory Year, 1970–1971. Volume II. XI, 786 pages. 1972. DM 40,–

Vol. 248: Lectures on the Applications of Sheaves to Ring Theory. Tulane University Ring and Operator Theory Year, 1970–1971. Volume III. VIII, 315 pages. 1971. DM 26,–

Vol. 249: Symposium on Algebraic Topology. Edited by P. J. Hilton. VII, 111 pages. 1971. DM 16,–

Vol. 250: B. Jónsson, Topics in Universal Algebra. VI, 220 pages. 1972. DM 20,–

Vol. 251: The Theory of Arithmetic Functions. Edited by A. A. Gioia and D. L. Goldsmith VI, 287 pages. 1972. DM 24,–

Vol. 252: D. A. Stone, Stratified Polyhedra. IX, 193 pages. 1972. DM 18,–

Vol. 253: V. Komkov, Optimal Control Theory for the Damping of Vibrations of Simple Elastic Systems. V, 240 pages. 1972. DM 20,–

Vol. 254: C. U. Jensen, Les Foncteurs Dérivés de $\varprojlim$ et leurs Applications en Théorie des Modules. V, 103 pages. 1972. DM 16,–

Vol. 255: Conference in Mathematical Logic – London '70. Edited by W. Hodges. VIII, 351 pages. 1972. DM 26,–

Vol. 256: C. A. Berenstein and M. A. Dostal, Analytically Uniform Spaces and their Applications to Convolution Equations. VII, 130 pages. 1972. DM 16,–

Vol. 257: R. B. Holmes, A Course on Optimization and Best Approximation. VIII, 233 pages. 1972. DM 20,–

Vol. 258: Séminaire de Probabilités VI. Edited by P. A. Meyer. VI, 253 pages. 1972. DM 22,–

Vol. 259: N. Moulis, Structures de Fredholm sur les Variétés Hilbertiennes. V, 123 pages. 1972. DM 16,–

Vol. 260: R. Godement and H. Jacquet, Zeta Functions of Simple Algebras. IX, 188 pages. 1972. DM 18,–

Vol. 261: A. Guichardet, Symmetric Hilbert Spaces and Related Topics. V, 197 pages. 1972. DM 18,–

Vol. 262: H. G. Zimmer, Computational Problems, Methods, and Results in Algebraic Number Theory. V, 103 pages. 1972. DM 16,–

Vol. 263: T. Parthasarathy, Selection Theorems and their Applications. VII, 101 pages. 1972. DM 16,–

Vol. 264: W. Messing, The Crystals Associated to Barsotti-Tate Groups: with Applications to Abelian Schemes. III, 190 pages. 1972. DM 18,–

Vol. 265: N. Saavedra Rivano, Catégories Tannakiennes. II, 418 pages. 1972. DM 26,–

Vol. 266: Conference on Harmonic Analysis. Edited by D. Gulick and R. L. Lipsman. VI, 323 pages. 1972. DM 24,–

Vol. 267: Numerische Lösung nichtlinearer partieller Differential- und Integro-Differentialgleichungen. Herausgegeben von R. Ansorge und W. Törnig, VI, 339 Seiten. 1972. DM 26,–

Vol. 268: C. G. Simader, On Dirichlet's Boundary Value Problem. IV, 238 pages. 1972. DM 20,–

Vol. 269: Théorie des Topos et Cohomologie Etale des Schémas. (SGA 4). Dirigé par M. Artin, A. Grothendieck et J. L. Verdier. XIX, 525 pages. 1972. DM 50,–

Vol. 270: Théorie des Topos et Cohomologie Etle des Schémas. Tome 2. (SGA 4). Dirigé par M. Artin, A. Grothendieck et J. L. Verdier. V, 418 pages. 1972. DM 50,–

Vol. 271: J. P. May, The Geometry of Iterated Loop Spaces. IX, 175 pages. 1972. DM 18,–

Vol. 272: K. R. Parthasarathy and K. Schmidt, Positive Definite Kernels, Continuous Tensor Products, and Central Limit Theorems of Probability Theory. VI, 107 pages. 1972. DM 16,–

Vol. 273: U. Seip, Kompakt erzeugte Vektorräume und Analysis. IX, 119 Seiten. 1972. DM 16,–

Vol. 274: Toposes, Algebraic Geometry and Logic. Edited by. F. W. Lawvere. VI, 189 pages. 1972. DM 18,–

Vol. 275: Séminaire Pierre Lelong (Analyse) Année 1970–1971. VI, 181 pages. 1972. DM 18,–

Vol. 276: A. Borel, Représentations de Groupes Localement Compacts. V, 98 pages. 1972. DM 16,–

Vol. 277: Séminaire Banach. Edité par C. Houzel. VII, 229 pages. 1972. DM 20,–

Vol. 278: H. Jacquet, Automorphic Forms on GL(2). Part II. XIII, 142 pages. 1972. DM 16,–

Vol. 279: R. Bott, S. Gitler and I. M. James, Lectures on Algebraic and Differential Topology. V, 174 pages. 1972. DM 18,–

Vol. 280: Conference on the Theory of Ordinary and Partial Differential Equations. Edited by W. N. Everitt and B. D. Sleeman. XV, 367 pages. 1972. DM 26,–

Vol. 281: Coherence in Categories. Edited by S. Mac Lane. VII, 235 pages. 1972. DM 20,–

Vol. 282: W. Klingenberg und P. Flaschel, Riemannsche Hilbertmannigfaltigkeiten. Periodische Geodätische. VII, 211 Seiten. 1972. DM 20,–

Vol. 283: L. Illusie, Complexe Cotangent et Déformations II. VII, 304 pages. 1972. DM 24,–

Vol. 284: P. A. Meyer, Martingales and Stochastic Integrals I. VI, 89 pages. 1972. DM 16,–

Vol. 285: P. de la Harpe, Classical Banach-Lie Algebras and Banach-Lie Groups of Operators in Hilbert Space. III, 160 pages. 1972. DM 16,–

Vol. 286: S. Murakami, On Automorphisms of Siegel Domains. V, 95 pages. 1972. DM 16,–

Vol. 287: Hyperfunctions and Pseudo-Differential Equations. Edited by H. Komatsu. VII, 529 pages. 1973. DM 36,–

Vol. 288: Groupes de Monodromie en Géométrie Algébrique. (SGA 7 I). Dirigé par A. Grothendieck. IX, 523 pages. 1972. DM 50,–

Vol. 289: B. Fuglede, Finely Harmonic Functions. III, 188. 1972. DM 18,–

Vol. 290: D. B. Zagier, Equivariant Pontrjagin Classes and Applications to Orbit Spaces. IX, 130 pages. 1972. DM 16,–

Vol. 291: P. Orlik, Seifert Manifolds. VIII, 155 pages. 1972. DM 16,–

Vol. 292: W. D. Wallis, A. P. Street and J. S. Wallis, Combinatorics: Room Squares, Sum-Free Sets, Hadamard Matrices. V, 508 pages. 1972. DM 50,–

Vol. 293: R. A. DeVore, The Approximation of Continuous Functions by Positive Linear Operators. VIII, 289 pages. 1972. DM 24,–.

Vol. 294: Stability of Stochastic Dynamical Systems. Edited by R. F. Curtain. IX, 332 pages. 1972. DM 26,–

Vol. 295: C. Dellacherie, Ensembles Analytiques, Capacités, Mesures de Hausdorff. XII, 123 pages. 1972. DM 16,–

Vol. 296: Probability and Information Theory II. Edited by M. Behara, K. Krickeberg and J. Wolfowitz. V, 223 pages. 1973. DM 20,–

Vol. 297: J. Garnett, Analytic Capacity and Measure. IV, 138 pages. 1972. DM 16,–

Vol. 298: Proceedings of the Second Conference on Compact Transformation Groups. Part 1. XIII, 453 pages. 1972. DM 32,–

Vol. 299: Proceedings of the Second Conference on Compact Transformation Groups. Part 2. XIV, 327 pages. 1972. DM 26,–

Vol. 300: P. Eymard, Moyennes Invariantes et Représentations Unitaires. II. 113 pages. 1972. DM 16,–

Vol. 301: F. Pittnauer, Vorlesungen über asymptotische Reihen. VI, 186 Seiten. 1972. DM 18,–

Vol. 302: M. Demazure, Lectures on p-Divisible Groups. V, 98 pages. 1972. DM 16,–

Vol. 303: Graph Theory and Applications. Edited by Y. Alavi, D. R. Lick and A. T. White. IX, 329 pages. 1972. DM 26,–

Vol. 304: A. K. Bousfield and D. M. Kan, Homotopy Limits, Completions and Localizations. V, 348 pages. 1972. DM 26,–

Vol. 305: Théorie des Topos et Cohomologie Etale des Schémas. Tome 3. (SGA 4). Dirigé par M. Artin, A. Grothendieck et J. L. Verdier. VI, 640 pages. 1973. DM 50,–

Vol. 306: H. Luckhardt, Extensional Gödel Functional Interpretation. VI, 161 pages. 1973. DM 18,–

Vol. 307: J. L. Bretagnolle, S. D. Chatterji et P.-A. Meyer, Ecole d'été de Probabilités: Processus Stochastiques. VI, 198 pages. 1973. DM 20,–

Vol. 308: D. Knutson, λ-Rings and the Representation Theory of the Symmetric Group. IV, 203 pages. 1973. DM 20,–

Vol. 309: D. H. Sattinger, Topics in Stability and Bifurcation Theory. VI, 190 pages. 1973. DM 18,–

Vol. 310: B. Iversen, Generic Local Structure of the Morphisms in Commutative Algebra. IV, 108 pages. 1973. DM 16,–

Vol. 311: Conference on Commutative Algebra. Edited by J. W. Brewer and E. A. Rutter. VII, 251 pages. 1973. DM 22,–

Vol. 312: Symposium on Ordinary Differential Equations. Edited by W. A. Harris, Jr. and Y. Sibuya. VIII, 204 pages. 1973. DM 22,–

Vol. 313: K. Jörgens and J. Weidmann, Spectral Properties of Hamiltonian Operators. III, 140 pages. 1973. DM 16,–

Vol. 314: M. Deuring, Lectures on the Theory of Algebraic Functions of One Variable. VI, 151 pages. 1973. DM 16,–

Vol. 315: K. Bichteler, Integration Theory (with Special Attention to Vector Measures). VI, 357 pages. 1973. DM 26,–

Vol. 316: Symposium on Non-Well-Posed Problems and Logarithmic Convexity. Edited by R. J. Knops. V, 176 pages. 1973. DM 18,–

Vol. 317: Séminaire Bourbaki – vol. 1971/72. Exposés 400–417. IV, 361 pages. 1973. DM 26,–

Vol. 318: Recent Advances in Topological Dynamics. Edited by A. Beck, VIII, 285 pages. 1973. DM 24,–